# CAMBRIDGE EARTH SCIENCE SERIES

# Earthquake mechanics

## IN THIS SERIES

# Earthquake mechanics

## K. KASAHARA

PROFESSOR, EARTHQUAKE RESEARCH INSTITUTE
UNIVERSITY OF TOKYO

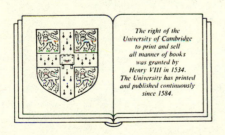

The right of the
University of Cambridge
to print and sell
all manner of books
was granted by
Henry VIII in 1534.
The University has printed
and published continuously
since 1584.

## CAMBRIDGE UNIVERSITY PRESS

CAMBRIDGE
LONDON  NEW YORK  NEW ROCHELLE
MELBOURNE  SYDNEY

Published by the Press Syndicate of the University of Cambridge
The Pitt Building, Trumpington Street, Cambridge CB2 1RP
32 East 57th Street, New York, NY 10022, USA
10 Stamford Road, Oakleigh, Melbourne 3166, Australia

First published 1981
Reprinted 1985

Printed in Great Britain at the University Press, Cambridge


*British Library Cataloguing in Publication Data*

Kasahara, K.
Earthquake mechanics
(Cambridge earth science series)
1. Earthquakes   I. Title   II. Series
551.2'2   QE534.2   80-40055
ISBN 0 521 22736 4

# Contents

# Preface

One of the most important insights gained by modern seismology is the recognition that earthquakes, especially shallow ones, are undoubtedly caused by faults in the Earth. This conclusion has been reached through a long process of adaptation and discarding of theories since the birth of seismology. The convergence of hypotheses to form one unified theory is indeed the basis of modern seismology, enabling substantial advances to be achieved since the 1960s.

The improvement in the logical structure of seismological theory has been accompanied by significant advances in experimental technology, which have yielded new observations leading to advanced theories and numerical techniques, as well as forging close links with various related disciplines in the solid-earth sciences. Briefly, progress in contemporary seismology must be viewed, in my opinion, from two contrasting standpoints: convergence in the logical background and divergence in experimental technology and instrumentation.

As a volume in the Cambridge Earth Science Series, this book aims primarily to introduce the reader to our current understanding of earthquakes, with greatest stress on the basic concepts upon which all progress is, or has been, founded. It is outside the scope of this book to discuss specialized problems in detail. Most chapters are based on lectures which I have given in the past decade to beginning graduate students at the several universities with which I have been associated. Consequently this book assumes a familiarity with the basic concepts of seismology and geophysics.

As will be seen from the contents list, subjects are developed more or less in a logical order, starting from the preliminary study of an earthquake source, and then proceeding to further specific problems such as static, dynamic, physical and tectonic aspects of earthquake sources and their environments in the Earth. The final chapter briefly discusses problems in earthquake prediction, a vast new frontier in our research field which must be explored with all our resources.

Because Système International (SI) units are still not widely used in this field, I have obtained permission from the Cambridge University Press to use cgs units in this book. A table showing conversions from cgs to SI units is given on page xiii.

The expression log in the text and equations denotes a common logarithm, unless otherwise specially remarked.

The type of magnitude of each earthquake quoted in the text is taken from the relevant reference, but it is generally understood that those for the Japanese earthquakes refer to the Rika-nenpyo system (*Annual Table of Scientific Constants,* Maruzen Publishing Co., Tokyo).

I acknowledge with thanks the invitation of Professor A. H. Cook, FRS, of Cambridge University to contribute to the Cambridge Earth Science Series, and his continuous support during the course of publishing. I also wish to express my thanks to colleagues at the Earthquake Research Institute including Dr T. Maruyama, Dr K. R. Rybicki (on leave from the Institute of Geophysics, Polish Academy of Sciences), Professor C. H. Scholz (on leave from the Lamont-Doherty Geological Observatory, Columbia University) and Dr W. Thatcher (on leave from the US Geological Survey), as well as to Professor T. Rikitake, Tokyo Institute of Technology, for their critical reading of this manuscript and for valuable advice. In particular, §4.3.2. owes much to the cooperation of Dr Maruyama. I am also much obliged to Dr E. R. Lapwood, Emeritus Reader at Cambridge University, who carefully reviewed the manuscript and gave detailed suggestions for improvement during his stay at this Institute.

I should also like to express my thanks to the staff of the Cambridge University Press, especially to Drs A. K. Parker and C. A. Lang, for the care taken in the publication of this book.

*Earthquake Research Institute*                    KEICHI KASAHARA
*University of Tokyo*
*February 1978*

# Table of conversions

| Physical quantity | cgs unit | SI unit (converted) |
|---|---|---|
| length | 1 cm | $10^{-2}$ m (metre) |
| mass | 1 g | $10^{-3}$ kg (kilogram) |
| time | 1 s | 1 s (second) |
| energy | 1 g cm$^2$ s$^{-2}$ (erg) | $10^{-7}$ J (joule, m$^2$kg s$^{-2}$) |
| force | 1 g cm s$^{-2}$ (dyn) | $10^{-5}$ N (newton, m kg s$^{-2}$) |
| pressure | 1 dyn cm$^{-2}$ ($10^{-6}$ bar) | $10^5$ Pa (pascal, m$^{-1}$ kg s$^{-2}$) |
| area | 1 cm$^2$ | $10^{-4}$ m$^2$ |
| volume | 1 cm$^3$ (cc) | $10^{-6}$ m$^3$ |
| density | 1 g cm$^{-3}$ | $10^3$ kg m$^{-3}$ |
| velocity | 1 cm s$^{-1}$ | $10^{-2}$ m s$^{-1}$ |
| acceleration | 1 cm s$^{-2}$ (gal) | $10^{-2}$ m s$^{-2}$ |
| moment of force | 1 dyn cm | $10^{-7}$ N m |
| viscosity | 1 dyn s cm$^{-2}$ (poise) | $10^{-1}$ Pa s |

# 1 Framework of seismology

## 1.1 EARTHQUAKE MECHANICS STUDY IN SEISMOLOGY

Seismology is primarily the study of earthquakes. The immutable desire of mankind to better understand the nature of earthquakes has sustained advance in seismology since its early days. With its increasing activity, the scope of seismology has been broadened considerably, as can be seen from the contents of any recent bulletin or journal in this field.

Fig. 1.1 shows the framework of seismology in terms of three principal groups of study. The first concerns the use of seismic waves as probes in the exploration of the Earth's interior. Theories and observations of wave propagation through the Earth and the interpretation of wave propagation in terms of the Earth's internal structures have been major problems in seismology since the last century. Studies of this type were later expanded to include further topics concerned with the Earth, thus augmenting our knowledge of its physical state, its history as a planet, the tectonic history on the surface, and so on.

The third group shown in fig. 1.1 is the application of seismological knowledge to human activities. Disaster prevention, seismic prospecting for natural resources, detection of underground nuclear explosions, etc.

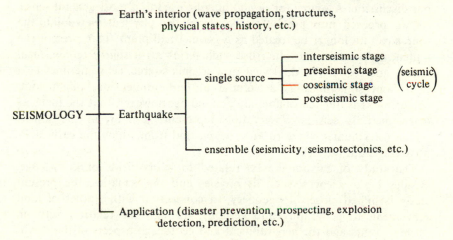

Fig. 1.1. Principal divisions of seismology.

1

are among these. Earthquake prediction may be added to this group, although it is inseparable from basic research on the nature of the earthquake itself.

The nature of earthquakes (the second group in fig. 1.1) is perhaps the most fundamental problem in seismology. This can be divided into two branches. The first branch deals with the general features of an earthquake, i.e. an individual earthquake source is analysed to give a general and comprehensive picture of an earthquake phenomenon. In the second branch, on the other hand, the activity of earthquakes as a group is the major interest. The mutual relation between earthquakes, and the dependence of earthquake occurrence on tectonic circumstances are examples of problems. Compared with the geographical and geological characteristics of this branch, the first one may be regarded to be more physical.

The second group of studies, i.e. that concerned with the mechanics of earthquakes, which is the title of this book, is primarily concerned with problems associated with a single source. In the following chapters, discussion will be developed mainly along this line. As occasion demands, however, the discussion may be extended to include further topics in order to build a broad foundation upon which to base an understanding of the nature of the earthquake.

A feature of an earthquake source may be expressed in various ways, depending upon which aspect of the phenomenon is of most concern. When we deal with a source of seismic waves macroscopically, a *focus*, or *hypocentre*, is used to specify a point in the Earth, from which waves are radiated (the *epicentre* is the vertical projection of the hypocentre on the surface). When we are concerned more with the physical mechanism, we may use a *fault* or *fracture* (or *rupture*), depending on the specification in our discussion. A *source* (or *origin*) may be used in a most general sense. As we proceed into precise studies from the physical viewpoint, the source can no longer be treated as a geometrical point. To represent the finiteness, therefore, we often use such terms as, a *source region, focal region*, and so on. The dual view of a seismic source, i.e. sometimes as a point and sometimes as a volume of finite dimension, might look confusing. It is our reasonable understanding, however, that the focus as determined by seismic observation represents a point in the source region (or volume) where rupture begins and from which the earliest P-wave starts out (§2.3.1).

The study of earthquakes is related to many fields of knowledge. Besides basic sciences, such as physics and mathematics, the present study is linked closely to geodesy, in connection with studies of land movement, to geography and geology, through the relation between seismic events and the geographical and geological aspects of the earthquake region, to rock mechanics, for a better understanding of fracture mechanisms, to engineering in connection with disaster prevention problems, and even to government policy if earthquake prediction is put into practice. Like many other fields of geoscience, the study of earthquakes is essentially an interdisciplinary science.

## 1.2 AN EARTHQUAKE MACHINE

The operation of an earthquake source is often compared to that of a machine which accumulates energy from a deeper source and converts part of it instantaneously into kinetic energy, i.e. seismic disturbances (Matuzawa, 1964). Fig. 1.2 illustrates schematically the structure of an earthquake machine (Kasahara, 1969).

An earthquake field, located at the centre of the figure, corresponds to the main part of the machine. Let $S$ denote its physical state immediately before an earthquake. Then, the earthquake occurrence may be represented by an abrupt change of $S$ into $S'$. Therefore, a correct understanding of the transition $S \rightarrow S'$ must be the primary concern in earthquake process studies. In principle, the transition must be specified with respect to all the physical quantities at each point of the earthquake source region. Some of them, such as pressure, $\sigma$, and temperature, $T$, are quantities of primary importance, and others such as deviatoric pressure $\Delta\sigma$, and dissipation factor $Q$, may be of secondary importance. In practice, however, we may omit from our discussion those quantities which are insensitively related to the earthquake process. Studies to determine which of them are essential are urgently needed.

The earthquake field is the recipient of energy supplied by a deeper source, and accumulates potential energy until it reaches a critical state. It is easy to imagine that the mode of energy supply is not that of action at a distance but of action in a medium, so that we need to know about the structure with respect to density, $\rho$, elasticity, $c$, anelasticity, $\eta$, electrical properties $r$, etc. Also, we need to understand the background of the accumulation process, that is the mutual interaction between adjacent parts of the medium, and the tectonic history of the region.

When the physical state of the region fits certain criteria, a rupture will occur causing transition of the state from $S$ to $S'$. A critical factor in this process is the presence of triggering forces of external or internal origin. Once a catastrophic rupture occurs, it may trigger another rupture in the same region, as illustrated by a feedback loop in fig. 1.2. This process will continue during the postseismic period, until the region attains a new equilibrium.

This series of events is believed to occur repeatedly in a region. In other words, the seismic history of a region is schematized by repetition of a *seismic cycle*, which comprises the *interseismic*, *coseismic*, and *postseismic* stages in that order. A *preseismic* stage might be hypothesized

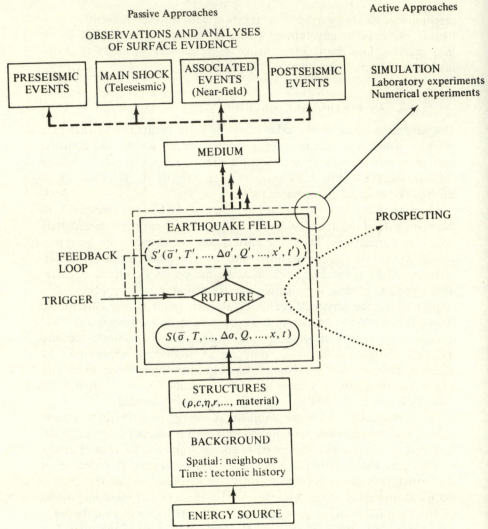

Fig. 1.2. An earthquake machine and various approaches to its mechanics. (From Kasahara, 1969.)

prior to the coseismic stage as discussed in §7.2.2. Fig. 1.3 illustrates the cycle in a region and gives the main problems associated with each stage.

The interseismic stage is characterized by accumulation of potential energy as stated previously. Energy source(s), the accumulation mechanism including its time rate and spatial distribution, and precursory changes in the physical state must be clarified for an understanding of the mechanism of this stage.

| INTERSEISMIC | PRESEISMIC | COSEISMIC | POSTSEISMIC |
|---|---|---|---|
| *Accumulation of potential energy* | *Anelastic behaviour of medium at the critical stress level* | *Conversion of potential energy to kinetic energy* | *Transition to new equilibrium* |
| Problems: | Problems: | Problems: | Problems: |
| Energy source(s) Accumulation mechanism, time rate, spatial distribution Associated changes in physical state | Recognition of seismic precursors and their mechanisms | Triggering mechanism, previous catastrophes Rupture: its type, geometry, development (velocity), wave radiation, stopping mechanism Changes in state and properties at focal region | Mechanism of aftershocks and other postseismic events Recovery of the changed state and properties Hysteresis |

Energy partition (efficiency, loss, interaction with neighbours)

Tectonic history

Fig. 1.3. Stages in a seismic cycle and various problems associated with them. (After Kasahara, 1969.)

Next comes the coseismic stage, or the stage of principal rupture, which is characterized by conversion of potential energy into kinetic energy, and might be preceded by the transitional preseismic stage. There are a number of problems related to this stage, but the essential ones are: (*a*) the triggering mechanism, including the influences of faulting in past catastrophes; (*b*) the rupture mechanism, including type, geometry, development (velocity of rupture propagation), wave radiation, and the stopping mechanism of the rupture; and (*c*) changes in state and properties at the focal region.

The transition to a new equilibrium will occur in the postseismic stage. Mechanisms of aftershocks and other postseismic events, recovery and hysteresis of physical conditions in the source region are examples of problems. This is the last of the three stages, but it only means the end of one seismic cycle. Tectonic aspects and the history of earthquakes suggest that there is a cyclic repetition of seismic events of similar type in a region. The end of a postseismic stage is also the beginning of the interseismic stage in the next cycle.

Balance and partition of energy are the most fundamental problems in the physics of focal processes. Efficiency of energy conversion, energy loss in the various stages, and interaction with the adjacent regions are problems in all three stages. The tectonic implication of seismic events is also a notable problem.

Let us return to fig. 1.2, which shows several possible approaches to the goal. As our observations are generally restricted to the Earth's surface, most of the signals from the source are obtained indirectly, i.e. through the medium responding to it. Modelling plays an important role in our study, as it allows us to make maximum use of the limited and indirect information available about the source. We construct a working hypothesis and improve it, in a trial and error manner, until we arrive at the most reasonable fit of the model to the observations.

From a methodological point of view, the interpretative work may be categorized as a passive approach, since we have no control on the signal source (top left of fig. 1.2). Alternative types, i.e. active approaches, are also possible (top right of fig. 1.2). Let us take a rupture experiment in rock mechanics as an example. If a laboratory experiment is planned carefully, it is possible to simulate plausible physical conditions in the focal region, so that some of the fundamental processes of rupturing at a given depth may be tested under the controlled conditions. Numerical experiments using a computer may be added to this group. Several workers have attempted simulations of this type and, although the models so far tested have been simple and elementary, there seems to be great potential in this approach. Seismic and geophysical prospecting across a seismic region by use of controlled signal sources has also shown great promise. Seismologists have renewed their interest in this technique, especially in its use in earthquake prediction research. Series of explosion-seismology experiments, carried out in several areas of high seismic risk, are undertaken to monitor temporal changes in seismic wave velocity (or the ratio $v_P/v_S$) which might occur as precursors to a violent earthquake in the region. Signals from natural earthquakes may also be used for this purpose if they occur at the proper localities.

## 1.3   BROAD-BAND SEISMOLOGY IN CRUST DYNAMICS

The frequency (or period) range of seismology has been broadened surprisingly in the last two or three decades. This is particularly apparent from a comparison of the instrumentation used in contemporary seismology with that used earlier this century.

In the earlier studies in seismology, scientists were concerned chiefly with the range of 'ordinary' vibration periods from one-tenth of a second to several seconds. This is understandable, since periods of seismic body waves lie predominantly in this range, and, in addition, seismic disturbances of these vibration periods have the most serious effect on buildings, and are those to which human bodies are most sensitive. In other words, we recognize 'ordinary earthquakes' by receiving a train of ground vibrations in this period range. The instrumentational consideration that recording of signals is relatively easy in this period range was another relevant factor.

New research fields in seismology have been introduced since the late 1940s, which have brought about a remarkable extension of the seismological band toward short and long periods. Development of high-gain seismographs has enabled us to discover micro- and ultra-microearthquakes, which are characterized by extremely low energy and extremely short period, far beyond the limit of ordinary earthquakes (fig. 1.4a).

Observations and analyses of long-period seismic waves have widened the band in the other direction. This is proving to be especially useful for earthquake mechanics studies, as it provides us with a new approach to the source mechanisms. Seismic signals recorded in the 100–1000 second range (fig. 1.4b) are particularly useful for this purpose. Sometimes periods of up to 3000 seconds are recorded, corresponding to the fundamental mode of free oscillation of the Earth.

In addition to these vibrating events, secular movements of the land have been included in the scope of seismology. The term *broad-band*, or *zero-frequency* seismology has been given to this aspect of research. Studies of the fault-origins of earthquakes and their tectonic implications have been developed by accumulating data of the Earth's deformation over a very long period of time, up to millions of years. Fig. 1.5 shows

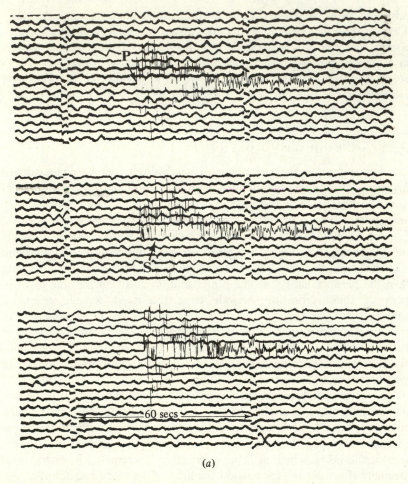

(a)

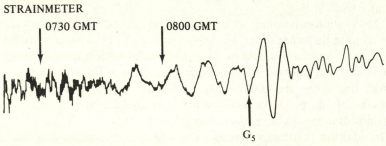

STRAINMETER

0730 GMT    0800 GMT

$G_5$

(b)

Fig. 1.4. Comparison of seismograms of very small (micro-) and very large earthquakes (station: Golden Colorado). (*a*) A local earthquake, northeast of Denver, Colorado (14 October 1966, $M = 2.4$) recorded on three components of a Benioff short-period seismograph (vertical, NS and EW, from top to bottom). Symbols P and S indicate arrivals of compressional and shear waves respectively. (*b*) The Alaskan earthquake (28 March 1964, $M = 8.5$) recorded on the NS-component of a strain seismograph. $G_s$ indicates surface waves that have travelled twice round the globe before being recorded. (From Simon, 1968. By courtesy of the Colorado School of Mines.)

the time ranges within which the various disciplines operate. It is divided into vibratory and secular movements, both being earthquake-associated crustal movements in the wide sense. There are four time bands introduced: seismological, geodetic, geomorphological, and geological. A

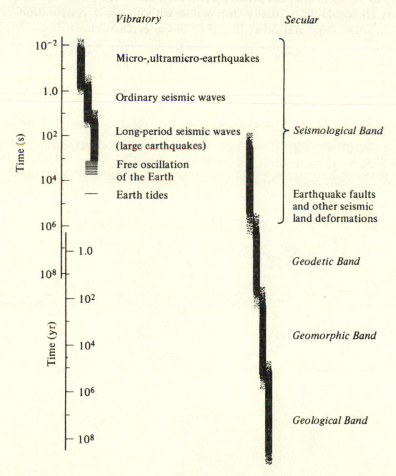

Fig. 1.5. Time ranges of the several disciplines related to broad-band seismology. (From Kasahara, 1971.)

historical and archaeological band could be included between the seismological and geodetic bands to cover the $10^2$–$10^5$ year interval.

Recent technical advances have made it possible to make geodetic measurements more accurately and more quickly than we could earlier. Yet, time factors still limit the geodetic approach. On the one hand, resurveys cannot in general be made more frequently than once a year, and on the other hand, the oldest available surveys are no more than 100 years old. Long-term movements must be studied by geomorphological and geological means, covering the time ranges of $10^2$–$10^5$ years, and $10^5$–$10^7$ (or more) years, respectively.

New research fields have been created by linking two neighbouring disciplines. Recent studies in earthquake mechanics are excellent examples of such interdisciplinary combination. Progress in this field of research would be virtually impossible without close cooperation between seismology and other branches of the earth-sciences.

# 2  Magnitude and volume of a source

## 2.1  EARTHQUAKE MAGNITUDE

Several measures of the size of an earthquake have been proposed. The magnitude scale is, without doubt, the most successful among them.

The earthquake magnitude is basically a relative scale. It defines a standard size of earthquake and rates the others in a relative manner by their maximum amplitude under identical observational conditions. This is evident from Richter's (1935, 1958) definition:

$$M = \log \left[ A(\Delta)/A_0(\Delta) \right] = \log A(\Delta) - \log A_0(\Delta), \qquad (2.1)$$

where $\Delta$ is an epicentral distance, and $A_0$ and $A$ denote the maximum trace amplitudes, on a specified seismograph, of the standard event and of the one to be measured, respectively. The standard earthquake, i.e. $M = 0$ ($= \log 1$) in Richter's formula, is such as to give the maximum trace amplitude of 1 $\mu$m on a Wood–Anderson type seismograph at $\Delta = 100$ km.

For local earthquakes in California, an empirical formula (see fig. 2.1) was obtained by Richter (1935):

$$\log A_0 = 6.37 - 3 \log \Delta, \qquad (2.2)$$

where $A_0$ is measured in $\mu$m, $\Delta$ in km and $\Delta \leqq 600$ km. The magnification of a Wood–Anderson seismograph is $\times 2800$, so we may write

$$\log A = \log (2800\, a), \qquad (2.3)$$

replacing the maximum trace amplitude, $A$, by the ground amplitude, $a(\mu$m). Substituting (2.2) and (2.3) into (2.1) we obtain

$$M = \log a + 3 \log \Delta - 2.92. \qquad (2.4)$$

Equation (2.4) is more generally applicable than (2.1), as it may be used for any type of seismograph, providing the correct value of the ground amplitude, $a$, is known.

Further modifications have since been introduced to the magnitude definition, so that it is now possible to measure the size of a distant or deep earthquake. The magnitudes used for these purposes are the *surface wave magnitude* ($M_s$) and the *body wave magnitude* ($m_b$). In order to

11

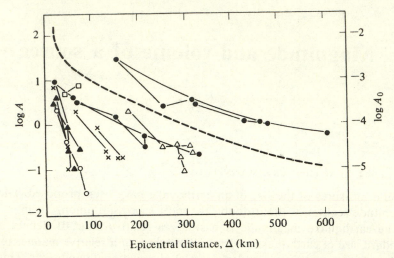

Fig. 2.1. Logarithm of the maximum trace amplitude, $A$ (where $A$ is measured in mm), versus epicentral distance, $\Delta$ (in km), for local earthquakes in California. The broken line shows the maximum trace amplitude of the standard-size event, $A_0$ (scale given at right of figure) against $\Delta$. (After Richter, 1958, *Elementary Seismology*, W. H. Freeman and Company. Copyright © 1958.)

distinguish these from the magnitudes used in the original definition, the magnitude $M$ in (2.1) or (2.4) is sometimes called the Richter *local magnitude* ($M_L$). The latter magnitude may only be used for shallow local earthquakes ($\Delta \leq 600$ km), such as the type studied by Richter.

For $\Delta > 600$ km, seismograms are dominated by surface waves of periods about 20 s rather than S body waves. Consequently, the maximum amplitude, $a$, depends on the distance, $\Delta$, in a manner different from that given in (2.2) and (2.4). Thus, the surface wave magnitude, $M_S$, has been introduced for a teleseismic case as

$$M_S = \log a + \alpha \log \Delta + \beta, \qquad (2.5)$$

where $a(\mu m)$ represents the maximum amplitude of the horizontal ground displacement for surface waves of about 20 s period. The constants $\alpha$ and $\beta$ are determined empirically by considering several reference earthquakes of known magnitude, so that (2.5) gives values consistent with those obtained by application of (2.1) and (2.4). In practice, the constants determined by various researchers scatter about the original data of Gutenberg (1945) ($\alpha = 1.656$, $\beta = 1.818 + C$ where C is a station constant), reflecting the effect of local conditions, and where $\Delta$ and $a$ are given in km and $\mu m$ respectively.

Equations (2.1) to (2.5) apply only to shallow earthquakes, with depths less than approximately 60 km. The surface waves produced by deeper earthquakes do not show the same relation between log $a$ and $\Delta$ as the shallow earthquakes, and therefore the above formulae are not applicable to this case. Gutenberg & Richter (1956*a*) conducted a laborious investigation of the variation of P, S, and other phases of body waves with distance for shallow and deep earthquakes. On the basis of these data, they proposed a body wave magnitude $m_b$ using $\log(a/T)$ in place of $\log a$ in the previous formulae. That is,

$$m_b = \log(a/T) + Q(h, \Delta), \tag{2.6}$$

where $T$ is the period, in seconds, of the measured wave, and $a$ represents ground amplitude in $\mu$m computed from the trace measurement and the constants of the particular instrument used. $Q(h, \Delta)$ is a parameter which is given empirically and is a function of the depth, $h$, of the seismic origin and the epicentral distance $\Delta$. The following formulae have been found empirically:

$$m_b = 2.5 + 0.63\, M, \tag{2.7}$$

or,

$$M = 1.59\, m_b - 3.97. \tag{2.8}$$

Note that the two values agree at $m_b = M \approx 6\frac{3}{4}$; above this $M > m_b$, below it $M < m_b$. Equation (2.6) implies that there is a connection between $m_b$ and the velocity of the ground motion.

With continuing expansion of its original concept, the earthquake magnitude scale has become the standard fundamental parameter with which to quantify an earthquake. No other measure represents the overall size of the earthquake quite as simply and concisely. To ensure that this convenient scale is used appropriately, however, the following points should be noted.

Formulae for magnitude determination are empirically derived from observation after radical simplification of the complex physical process at the seismic source. Consequently, it is generally understood that a magnitude determination may involve uncertainties of between 0.2 and 0.3, even under the most favourable conditions.

The magnitude scale was originally introduced for local events. Definition of the surface wave magnitude has enabled us to measure larger earthquakes at teleseismic distances, but $M_S$ tends to saturate for very large earthquakes beyond magnitude 8. Recently, a new magnitude scale $M_W$ has been defined in order to meet the demand for proper quantification of such very large earthquakes (§6.1.3).

Similar but independent scales have been used in several countries. The Russian *energy index*, $K$, is defined as the common logarithm of the seismic energy produced in an earthquake, given in joules (J). For example, $K = 16$ corresponds to a seismic energy of $10^{16}$ J, i.e. $10^{23}$ ergs, which roughly corresponds to an $M = 7.5$ earthquake. $M_K$, which was proposed by Kawasumi, is another example. It is defined as the seismic intensity on the Japan Meteorological Agency (JMA) scale felt at an epicentral distance of 100 km. It is especially useful for scaling of historical earthquakes, the seismic intensity of which can often be determined retrospectively by examination of the spatial distribution of earthquake effects. Empirically, this scale tends to give a numerical figure approximately 0.5 higher than the conventional magnitude.

### 2.2    SEISMIC ENERGY

As noted in §2.1, the magnitude scale measures earthquake size in a relative manner. In other words, it compares large and small earthquakes quantitatively, but indicates little about the physical properties of their sources. For a more precise discussion of a seismic source property, therefore, we need to relate the scale to a basic physical parameter such as energy.

Let us introduce a point source radiating a wave train uniformly in all directions (fig. 2.2). Suppose a seismic wave reaching a station at the epicentre is given in terms of the ground displacement as

$$x = a_0 \cos (2\pi t/T_0), \tag{2.9}$$

then the velocity of the ground motion there is

$$v = -(2\pi a_0/T_0) \sin (2\pi t/T_0), \tag{2.10}$$

where $a_0$ and $T_0$ denote the amplitude and period of the wave respectively. Hence the density of the kinetic energy of the ground motion, $e$ (per unit volume) is

$$e = (\rho/2T_0) \int_0^{T_0} v^2 \mathrm{d}t = (\rho/2T_0)(2\pi a_0/T_0)^2 \int_0^{T_0} \sin^2 (2\pi t/T_0)\mathrm{d}t \tag{2.11}$$

$$= (\rho/4)(2\pi a_0/T_0)^2,$$

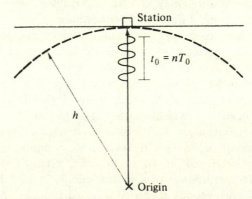

Fig. 2.2. Schematic diagram of part of a wave train from a point source approaching a station at the epicentre.

where $\rho$ is the density of the medium. If the wave train, of duration $t_0$, has $n$ wave periods in it $(t_0 = nT_0)$ and is propagated with velocity $c$; then the energy flow per unit area at the station is $ct_0 e$ (ignoring the effect of surface reflection). Therefore, integration over a spherical surface with radius $h$ (where $h$ is the depth of the origin), gives the total kinetic energy from the origin

$$E_k = 4\pi h^2 ct_0 e = 4\pi^3 h^2 ct_0 \rho (a_0/T_0)^2. \qquad (2.12)$$

The following factors must be considered to calculate the total seismic energy $E$:

(a) Since the mean potential and kinetic energies are equal, we may take $E = 2E_k$.

(b) Due to doubling of amplitude at the epicentre (free surface), $a_0 = 2a$, where $a_0$ denotes the amplitude recorded at the free surface.

(c) The calculation above dealt with the waves of maximum energy, which at short distances are S-waves. The energy of P-waves must be added, which is provisionally assumed to be half that of the S-waves (Gutenberg & Richter, 1956a).

Accordingly we obtain

$$E = 3\pi^3 h^2 ct_0 \rho (a_0/T_0)^2. \qquad (2.13)$$

Taking $c = 3.4$ km s$^{-1}$, $\rho = 2.7$ g cm$^{-3}$, and $h = 16$ km (a probable focal depth in southern California), (2.13) reduces to

$$\log E = 12.34 + 2 \log (a_0/T_0) + \log t_0. \qquad (2.14)$$

Gutenberg & Richter obtained the following empirical relations for earthquakes in southern California,

$$\log t_0 = -1 + 0.4 \log (a_0/T_0), \qquad (2.15)$$

$$\log (a_0/T_0) = m_b - 2.3, \qquad (2.16)$$

therefore,

$$\log E = 5.8 + 2.4 \, m_b, \qquad (2.17)$$

where $m_b$ is the body wave magnitude discussed in the previous section. Substituting (2.7) into (2.17), we obtain the magnitude–energy relation as

$$\log E = 11.8 + 1.5 \, M, \qquad (2.18)$$

which will be the relation referred to in the following chapters, unless otherwise stated (see also table 2.1 for numerical examples).

If $M$ is increased by 1.0 in (2.18), $E$ is magnified by a factor of $10^{1.5}$, i.e. approximately 32. In other words, the seismic energy of an $M = 6$ earthquake is about 32 times as large as that of an $M = 5$ earthquake, and is about 1000 times that of an $M = 4$ earthquake.

Energy is a well-defined physical quantity, but this does not mean that the estimation of the seismic energy, as introduced above, is of high precision. Many assumptions have been made in the derivation of the kinetic energy of waves, e.g. the wave form has been greatly simplified, we have ignored the azimuthal effect of wave radiation, attenuation during propagation, and wave behaviour near the surface, etc. Even if these factors are taken into account to determine an accurate measurement of the kinetic energy, the more essential question of whether or not this particular quantity can accurately represent all the energy processes at a seismic source must still be considered.

Table 2.1. *Magnitude and energy of earthquakes.*

| Magnitude, $M$ | Energy (erg) |
|---|---|
| 8.5 | $3.6 \times 10^{24}$ |
| 8.0 | $6.3 \times 10^{23}$ |
| 7.5 | $1.1 \times 10^{23}$ |
| 7.0 | $2.0 \times 10^{22}$ |
| 6.5 | $3.6 \times 10^{21}$ |
| 6.0 | $6.3 \times 10^{20}$ |
| 5.5 | $1.1 \times 10^{20}$ |
| 5.0 | $2.0 \times 10^{19}$ |
| 4.5 | $3.6 \times 10^{18}$ |
| 4.0 | $6.3 \times 10^{17}$ |

## 2.3   SOURCE VOLUME

### 2.3.1   *Dual view of an earthquake source*

The simplest and most conventional picture of an earthquake source is a point origin. In seismometry, for instance, we take this idea and locate an origin of seismic waves at a point in the Earth. The point-source model works very satisfactorily as long as our discussion is concerned with a macroscopic view of an earthquake phenomenon. For a more detailed discussion, however, the picture of a point source seems unrealistic.

Table 2.1 shows that a considerable amount of kinetic energy is released from even a small earthquake. If the earthquake is large the energy from the source will be vast. On physical grounds, it is hard to believe that a single point in rock can accumulate such a great amount of energy and radiate it instantaneously. Thus, we are naturally led to a hypothesis that the energetic processes at a source occupy a finite volume in the Earth.

There is a great deal of evidence which lends support to this hypothesis. Empirical formulae based on various data consistently indicate that the source dimensions are related to the magnitude. The larger the earthquake magnitude, the greater the volume of the source. Tentative calculations show that a large earthquake ($M \approx 8$) has a source volume of millions of cubic kilometres (cf. §2.3.5).

The above statement might lead to confusion about the dual view of a source – as a point source in the seismometrical sense, and, as a volume source in the energetic sense. In the case of a small earthquake, or in the case of a macroscopic study of a seismic event this dual approach causes little confusion. In other cases, however, an essential question about the definition of a source is raised. A reasonable explanation will be provided later (chapter 5) on the basis of the fault-origin model of earthquakes. Briefly, the source volume represents a space around a fault from which the strain energy is effectively released. Its dimension is comparable to that of the fault. The focus (point source) in the seismometrical sense is the point in this volume where the rupture begins and from which the earliest P-wave radiates. Therefore, the two views of a source are not contradictory, but complement each other when describing physical processes at the source.

### 2.3.2  Geometrical aspects

*Aftershock area.* Fig. 2.3 compares the aftershock distribution of two shallow earthquakes in Japan. There is a clear difference between the

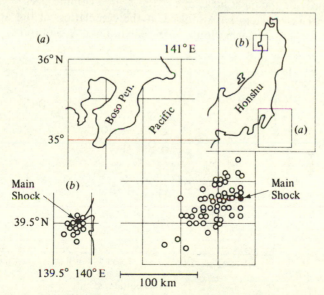

Fig. 2.3. Aftershock areas for earthquakes of different magnitudes: (*a*) off the Boso Peninsula, Honshu, Japan (1953, $M = 7.5$); and (*b*) on the Oga Peninsula, Honshu, Japan (1939, $M = 7.0$). (After Utsu & Seki, 1955.)

size and distribution of the aftershocks for these events. In the event shown in fig. 2.3*b* of magnitude $M = 7.0$, aftershocks are located in a limited area around the epicentre, whereas in the event shown in fig. 2.3*a* of magnitude $M = 7.5$, they occupy a much wider area. Quantitatively, the approximate sizes of their aftershock areas are 10 000 km² (fig. 2.3*a*) and 500 km² (fig. 2.3*b*). Utsu & Seki (1955) studied major earthquakes in the Japanese area and plotted their aftershock areas, *A*, against their magnitudes, *M* (fig. 2.4). They found an empirical formula for the relation between *A* and *M*:

$$\log A = 1.02 \, M + 6.0, \tag{2.19}$$

where *A* is measured in cm².

The mechanism of aftershock occurrence is unknown in detail, but it is undoubtedly associated with the main fracture at the source. If an aftershock area is indicative of the size of the main fracture, then the relation between *A* and *M* in (2.19) is strong evidence for the previous hypothesis concerning an earthquake source volume.

*Surface evidence.* Large shallow earthquakes sometimes produce remarkable surface effects such as fault offsets and land deformations. These provide the most measures of the source volume size.

Several researchers have looked at the correlation of the area of the coseismic land deformations with earthquake magnitude. Dambara

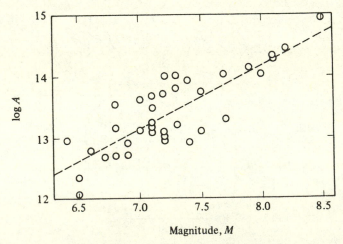

Fig. 2.4. Logarithm of aftershock area, *A* (where *A* is measured in cm²), versus magnitude, *M*, of the main shock. (After Utsu & Seki, 1955.)

(1966), for example, approximated the area as a circle with radius $r$ and found, on the basis of Japanese data, a formula which may be rewritten as

$$\log r = 0.51 \ M + 2.73, \tag{2.20}$$

where $r$ is measured in cm.

As pointed out by Tsuboi, (2.20) can be converted into an $A'$–$M$ relation ($A'$ being the area of land deformation) as follows:

$$\begin{aligned} \log A' &= \log \pi + 2 \log r \\ &= 1.02 \ M + 5.96, \end{aligned} \tag{2.21}$$

which suggests that $A' \approx A$ (cf. (2.19)). In other words, both aftershocks and land deformations tend to occupy approximately the same area around an epicentre. From (2.21), $A' \approx 100 \ \mathrm{km}^2$ for $M = 6$. In the case of the largest historical earthquake ($M = 8.6$), we obtain $A' \approx 50\,000 \ \mathrm{km}^2$ with effective radius $r \approx 120 \ \mathrm{km}$.

Statistical studies of earthquake faults by several researchers (Tocher, 1958; Iida, 1959, 1965; Press, 1967; and others) provide useful information for the present discussion. Although their conclusions are somewhat different, they may be reduced to an empirical formulae of the general form

$$\log L = p + qM, \tag{2.22}$$

where $L$ denotes the length of a fault, and $p$ and $q$ are constants (see §4.1.2 for further discussion).

The values of $p$ and $q$ show a degree of scatter reflecting, among other things, differences in regional structures. Yet, $q$ is generally found to fall within the range 0.5–1.2. Especially notable is the formula given by Otsuka (1965), which can be presented in the form

$$\log L_m = 3.2 + 0.5 \ M, \tag{2.23}$$

where $L_m$ (cm) denotes the upper limit of the fault length for a given magnitude. A value for $q$ of 0.5 is obtained, which is consistent with previous results (see (2.20)).

### 2.3.3 *Spectral aspects of seismic waves*

Many seismologists have shown that the period of the spectral peak for body and surface waves ($T$) increases with magnitude. For example, the

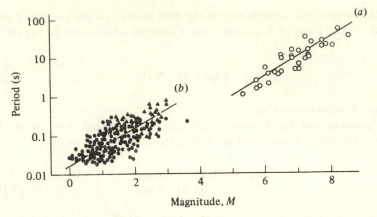

Fig. 2.5. Period of spectral peak (P-wave) versus magnitude: (*a*) for major shocks in Japan; and (*b*) for microearthquakes at Tsukuba Observatory, Japan. (After Terashima, 1968.)

following formula was given by Kasahara (1957) for P-wave spectra of large earthquakes ($M > 5$)

$$\log T = 0.51 \ M - 2.59. \tag{2.24}$$

For smaller earthquakes ($M < 3$), Terashima (1968) gives the relation

$$\log T = 0.47 \ M - 1.79. \tag{2.25}$$

Notice that the proportionality constants of $\log T$ to $M$ appear to be almost the same for the two equations. As shown in fig. 2.5, however, plots of magnitude versus period for the two groups do not meet, due to a considerable difference in the second terms in (2.24) and (2.25). Let us leave this problem for later discussion (chapter 5) and continue to study an earthquake volume.

A conventional explanation for the $T$–$M$ relation given above is based on a theory of wave radiation from a spherical source (Kasahara, 1957). Progress in the fault-origin model of earthquakes has yielded a more realistic explanation of the relation and will be discussed in §5.2.3.

### 2.3.4    *Ultimate strain and strain energy*

As with a brittle specimen under loading tests, the Earth's crust may break if it is strained beyond a certain limit. Fig. 2.6, taken from Tsuboi (1933), illustrates the coseismic strain field (shear) in the Tango earthquake (Honshu, Japan, 1927, $M = 7.5$) as derived from retriangulation data (see also fig. 4.5). Tsuboi noticed that, with the exception of the

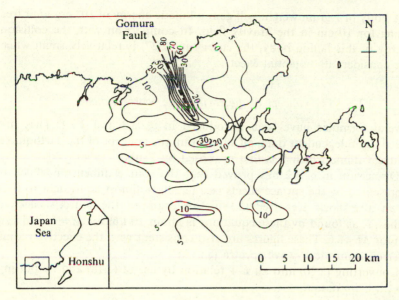

Fig. 2.6. Contours of equal maximum shear strain ($\times 10^5$) in the Tango earthquake (1927, $M = 7.5$) Honshu, Japan (see fig. 4.5 for the location of the fault). (After Tsuboi, 1933.)

faulted zones, crustal strains are of the order of $10^{-4}$ or less. He analysed the crustal strains associated with several seismic events and estimated the ultimate crustal strain as $1-2 \times 10^{-4}$ (Tsuboi, 1956); the Earth's crust may be strained up to this level elastically, but never beyond it without breaking (see Rikitake, 1976, for further data). This value seems extremely low when compared with data from laboratory experiments (which give a value of $10^{-3}$ or higher). This discrepancy is understandable as the actual crust is likely to contain numerous fractures and faults, which would be expected to reduce the macroscopic strength of the crust. Indeed, *a chain is no stronger than its weakest link*.

Let us refer to Tsuboi (1956) and estimate the strain energy in the crust under the critical condition. In the simplest case, in which the strain field is represented by a single component of shear, $\varepsilon$, the energy of the medium per unit volume is given by

$$e = \tfrac{1}{2}\mu\varepsilon^2, \tag{2.26}$$

where $\mu$ is the rigidity. Putting $\varepsilon = 1-2 \times 10^{-4}$ and $\mu = 5 \times 10^{11}$ dyn cm$^{-2}$ into (2.26), we obtain

$$e = 0.25-1.0 \times 10^4 \text{ erg cm}^{-3},$$

or,

$$e \approx 3000 \text{ erg cm}^{-3}. \tag{2.27}$$

A body of 1 gram weight will gain a kinetic energy of $10^4$ erg after free falling for 10 cm in the gravity field. In comparison with the collision energy of this falling body, the energy in (2.27) is relatively small when one considers its potential effect.

### 2.3.5    *The concept of the earthquake source volume*

Several formulae have been introduced in §§2.3.3 and 2.3.4. They are based on independent data, all supporting the concept of the earthquake source volume which was first proposed by Tsuboi (1956).

Discussion in §2.3.2 has proved that the source dimension which is represented by the surface effects (see (2.20)) is almost equivalent to that of an aftershock area (see (2.19)). For example, the source's effective radius, $r$, as found by these equations is about 10 km for $M=6$, and 120 km for $M=8.6$. These figures are also consistent with the effective radius inferred from the P-wave spectra (§2.3.3).

Converting (2.20) into an $E$–$V$ relation by use of (2.18) and by putting $V=\frac{4}{3}\pi r^3$, we obtain

$$\log E = \log V + 3.0, \tag{2.28}$$

or,

$$E/V = 1000 \text{ erg cm}^{-3}. \tag{2.29}$$

Notice that the energy density (2.29) has a value of the same order as $e$ given by (2.27). This suggests that:

(*a*) strain energy accumulated in the crust is a possible source of seismic energy;

(*b*) density of energy release from the source is basically independent of the magnitude, and is rather uniform, with a value of about $10^3$ erg cm$^{-3}$, which is similar to the potential energy in rocks at ultimate strain;

(*c*) consequently, the volume of strain release is the primary factor that specifies the quantity of seismic energy released and consequently, the magnitude of the resultant earthquake.

Therefore, we are led to the concept of the *earthquake source volume*, proposed by Tsuboi (1956).

### 2.4    EARTHQUAKE STATISTICS

The distribution of earthquakes in time and space is an elementary subject in seismology. Seismic observations have accumulated an extensive amount of information on seismic activity (*seismicity*) in various regions of the Earth and in various periods of time. Both earthquake frequency and energy release are well documented. In this book, however, these data will not be discussed in detail. For more information, the reader should refer to the extensive literature now available on this subject (e.g. Gutenberg & Richter, 1954; Richter, 1958; Lomnitz, 1974). The following represents only a small part of our current knowledge about seismicity, which is reviewed here for future reference.

### 2.4.1 *Earthquake ensemble and energy partition*

Table 2.2 presents the statistics of world seismicity. The mean annual frequency $N$ of shallow earthquakes is shown for classes of magnitude, $M$, from 3.0 to 8.9. Gutenberg and Richter (1944, 1954) proposed that the magnitude and frequency are related by

$$\log N = a + b(8 - M), \tag{2.30}$$

where $N$ denotes the frequency of events in a fixed time period and geographical region sorted into the magnitude class $M(\pm \Delta M)$. The data in table 2.2 (for $M \geq 6$) give values of $a$ and $b$ of $-0.48$ and $0.9$ respectively. Tests for various groups of earthquakes have proved that this is a very general formula and is widely applicable to seismic activity, although in each group the constant $a$ differs depending on the number of events in that group. The constant $b$ takes a value of about $0.9$ in most cases, which suggests that the relation between the relative frequency of earthquakes and their magnitude is similar for all groups, and that the frequency increases by a factor of about eight for each magnitude step down the scale. Despite several investigations, it is still not fully understood why the $b$-value tends to be about $0.9$, and more fundamentally, why energy release follows the partition law (2.30) for such a wide range of magnitudes. We may only say that the $b$-value seems to represent properties of the seismic medium in some respects. Significant deviations of $b$ from its standard value are occasionally observed, but only in special types of activity, such as swarm earthquakes and volcanic earthquakes.

Table 2.2. *Mean annual frequency of earthquake occurrence in the world. Values of* N *for* M $\geq 6$ *are taken from Gutenberg & Richter (1954). Values of* N *for* M $< 6$ *are extrapolated from Gutenberg & Richter's data using the magnitude–frequency relation (2.30).*

| Magnitude $M$ | $\geq 8$ | 7.9–7 | 6.9–6 | 5.9–5 | 4.9–4 | 3.9–3 |
|---|---|---|---|---|---|---|
| Frequency $N$ | 1 | 13 | 108 | 800 | 6200 | 49 000 |
| Energy ($10^{23}$ erg) | 13.7 | 12.0 | 1.1 | 0.8 | 0.2 | 0.05 |
| % of total energy released in one year | 49 | 43 | 4 | 3 | 1 | |

Table 2.3. *Numbers and energy of shocks in various regions for: (a) 1904–45 for M ≥ 7.8; (b) 1922–45, in general, for 7.7 ≥ M ≥ 7.0; (c) smaller earthquakes, M < 7.0. (After Gutenberg & Richter, 1954.)*

| Region | Total numbers | | | | Annual numbers | | | | | Energy in % | | |
|---|---|---|---|---|---|---|---|---|---|---|---|---|
| | Shallow | | Intermediate | Deep | Shallow | | | Intermediate | Deep | Shallow | Intermediate | Deep |
| | (a) | (b) | (a, b) | (a, b) | (a) | (b) | (c) | (a, b) | (a, b) | | | |
| Circum-Pacific | 74 | 284 | 120 | 31 | 1.75 | 10.14 | 86 | 4.08 | 1.08 | 75.4 | 89 | 100 |
| Mediterranean and trans-Asiatic | 16 | 30 | 12 | 0 | 0.38 | 1.07 | 10 | 0.39 | 0 | 22.9 | 11 | 0 |
| Others | 2 | 20 | 0 | 0 | 0.04 | 0.74 | 12 | 0 | 0 | 1.8 | 0 | 0 |
| Total | 92 | 334 | 132 | 31 | 2.2 | 11.9 | 108 | 4.47 | 1.08 | 100 | 100 | 100 |

After the constants, *a* and *b*, have been obtained for major earthquakes ($M \geq 6$), *N* for smaller earthquakes may be interpolated by use of (2.30). Then, by use of (2.18), *E* and, consequently, *NE* are calculated. Table 2.2 shows *NE* for the respective magnitude ranges. These data may be used for a study of the energy partition between large and small earthquakes. Notice that the majority of the total energy release is contributed by events of larger magnitude (93% from $M \geq 7$). Smaller earthquakes contribute little, in spite of their higher frequency of occurrence.

### 2.4.2 *Seismic geography*

Although earthquakes may occur anywhere on the Earth, the great majority are concentrated in the circum-Pacific belt. As shown in table 2.3, this area generates about 80% of the shallow shocks, 90% of the intermediate shocks (60–300 km deep), and nearly all the deeper ones. Considerable activity also occurs along the Mediterranean and trans-Asiatic belt.

Recent progress in seismometrical techniques has enabled us to study the global distribution in greater detail. We now recognize an almost linear alignment of epicentres running across the middle part of the oceanic areas (see Lomnitz, 1974; Rikitake, 1976). In the Atlantic Ocean, for example, a line of epicentres runs through its middle part from north to south, turning around the southern tip of Africa to enter the Indian Ocean. Then it divides into two branches, one reaching the Red Sea, and the other extending eastward to the south Pacific. At a point off the west coast of South America, the south Pacific branch splits again, one branch extending to the Chilean coast and the other running up to the Gulf of California. Unlike the major seismic belts previously discussed, these alignments are not the sites of extremely big earthquakes. Yet, they are of primary importance in the global tectonic system (chapter 7).

The statistics of the focal depths of earthquakes show that the majority of earthquakes, including those with the greatest energy, appear in the Earth's crust, or at a depth of 40 km or less. Deeper earthquakes occur with decreasing frequency down to the 250 km level. Below 250 km, down to 700 km, the vertical distribution of earthquakes is rather uniform, although, in some cases, there is a local increase of activity at a certain level (Gutenberg & Richter, 1954).

### 2.4.3 *Energy release in earthquake sequences*

Earthquake occurrence in an area looks irregular and random if the individual events in a sequence of earthquakes are considered. Yet, we notice several empirical rules which control the statistical aspects of earthquake sequences. The rule for aftershock frequency, as first pointed out by Omori (see Richter, 1958), is a famous example. Earthquake periodicity is another example of an important topic for which statistical tests have been made by many researchers. Let us leave a detailed discussion of these short-term aspects of earthquake sequences to other textbooks (e.g. Lomnitz, 1974), and turn our attention to the study of the long-term characteristics of seismic activity.

Fig. 2.7 illustrates the cumulative seismic energy released by major earthquakes ($M \geq 6$) in the Japanese area from 1884 to 1963. The curve is irregular, reflecting occasional high activity, yet it shows a general trend indicated by a pair of parallel lines $S$ and $S'$, which bound the upper and lower limits of fluctuation, respectively. Tsuboi derived a formula for $S$:

$$S = \Sigma E = (2.24 \ t + 1.91) \times 10^{23} \quad (\text{erg}), \tag{2.31}$$

(where $t$ is in years) together with an interpretation that the energy supply for potential earthquakes in this area is remarkably uniform with respect to time. On the basis of this finding, he hypothesized that one could estimate the maximum earthquake energy available at a particular time in a given area by considering the difference between the theoretical upper limit given by the $S$ curve and the cumulative energy released up to that time.

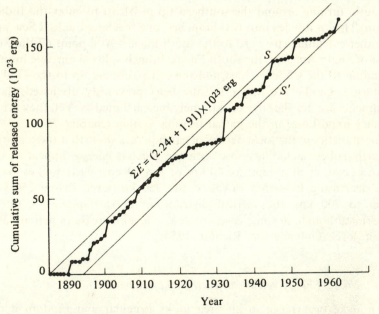

Fig. 2.7. Cumulative sum of energy released by earthquakes ($M \geq 6.0$) in and near Japan. (After Tsuboi, 1965.)

The question of whether the seismicity in an area is uniform, in a long-term sense, is a matter of concern in prediction research. If it is accepted that seismicity is uniform this would form a good base for statistical prediction. The concept of a seismicity gap, which is a useful working hypothesis for earthquake prediction, implicitly assumes uniform seismicity (see chapter 8).

# 3 Radiation pattern and focal mechanisms

## 3.1 RADIATION PATTERN

### 3.1.1 Observations

The polarity of the initial P-wave pulse from an earthquake will take either of two opposite senses, *compressional* (*pushed*) or *dilatational* (*pulled*), according to whether it is away from or toward the epicentre. The history of focal mechanism study may be traced back to the late 1910s, when Professor Shida of Kyoto University discovered a systematic distribution of the two senses of polarity in azimuth about the epicentre. The significance of this discovery for the study of source mechanisms is made clearer if we consider an explosive source, which generates compression in the P-phase in all directions.

There has been vigorous discussion about this problem, both from the experimental and theoretical sides, in order to determine laws of azimuthal distribution and to clarify the physical conditions that cause it. Progress in this field has been reviewed by many researchers, for example, Kawasumi (1937), Byerly (1955), Honda (1957, 1962), Bessonova *et al.* (1960), Balakina, Savarensky & Vvedenskaya (1961), Stauder (1962), Hodgson & Stevens (1964), and Stefánsson (1966). Proceedings of several symposia on this particular subject are also available, e.g. Hodgson (1959, 1961), Kasahara & Stevens (1969).

Focal mechanism studies were initially developed using local data. The seismic network in Japan was especially useful, in this respect, because stations are often well distributed about the epicentre. As an example, fig. 3.1 shows surface displacement in the initial P-phase caused by the earthquake of 7 March 1927 ($M = 7.5$) in the Tango district, Honshu, Japan (see also fig. 2.6 for the maximum shear field and fig. 4.5 for the horizontal land movements around the epicentre of this event). The systematic distribution of the initial P-pulse polarity in azimuth is evident in fig. 3.1, which shows the polarity and the amplitude of the initial P motion at each station as a vector. Stations to the east and west of the epicentre record dilatation, whereas compression is observed at stations to the south and north (though the data are sparse). The distribution of these polarities is so systematic in the figure that we may easily

28

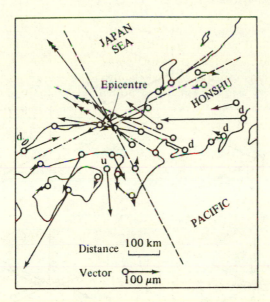

Fig. 3.1. Ground movements in the initial P-phase for the Tango earthquake, Honshu, Japan (1927, $M = 7.5$). Symbols u and d denote the upward and downward initial P-pulses respectively. (After Honda, 1957.)

draw a pair of lines, crossing at the epicentre, to separate the compressional and dilatational groups (see the dashed lines in the figure). Let us call these the *nodal lines* as the amplitude tends to go to zero along them. This type of distribution is sometimes called the *quadrant type* because of the perpendicular orientation of the two lines.

In the course of focal mechanism study, an alternative type of distribution (the so-called *cone-type*, see Kawasumi, 1937) was proposed to replace or complement the quadrant type. In the following discussion, however, we shall refer only to the quadrant type, as it is used widely in this research field.

### 3.1.2 *Force system at a point source*

The systematic distribution of the initial P-pulse polarity in azimuth has an essential bearing on the *focal mechanism*, i.e. the mechanism of wave radiation from a seismic focus. Theoretical models with various types of forces acting at a point source have been studied in order to explain the observations given above. The models shown in fig. 3.2c and e are especially interesting from the historical point of view (the model shown in fig. 3.2d is equivalent to that shown in fig. 3.2e in its seismic effect). As will be explained later (§6.1.2), we now accept the model shown in fig. 3.2e or its equivalent, fig. 3.2d, as a reasonable point-source model of

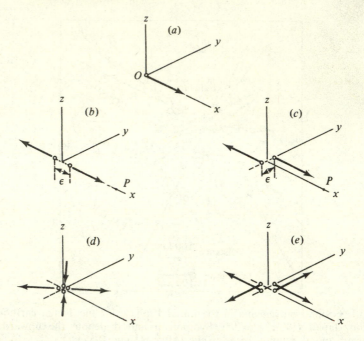

Fig. 3.2. Various force types at a point source: (*a*) a single force; (*b*) a pair of equal and opposite forces as a tension; (*c*) a pair of equal and opposite forces as a torque about the *z*-axis (type I); (*d*) two pairs of forces, tension and compression are of equal magnitude and perpendicular to each other (type II); (*e*) two pairs of forces. Their torques about the *z*-axis are of equal magnitude and are opposite to each other (type II).

earthquakes. The following discussion will be mainly concerned with this model, whereas the model shown in fig. 3.2*c* will be studied briefly for historical interest.

An earthquake is described, in a general sense, as an abrupt transition of the physical state in the Earth (§1.2). We may reasonably assume that the medium is initially in an elastic equilibrium. Suppose that a point in the medium is subject to a force $f(t)$, which rises from zero at $t = 0$, and reaches a certain level in a short time and remains constant thereafter. Then, we would observe transitional disturbances in the medium until it gains a new equilibrium.

Radiation of elastic waves from a point source has been studied theoretically since the middle of the last century. Love (1944), in his famous textbook on the theory of elasticity (first published in 1892), has given expressions like (3.1) for a displacement field due to a force acting at a point in an infinite medium. His theory was introduced into seismology by Nakano (1923), who founded the theory of seismic origins that many successors have applied to source mechanism studies (e.g.

Honda, 1957). Following Maruyama (1968), we will take a simpler view of the source mechanism than given in the pioneering studies mentioned above (see also the review papers listed in §3.1.1).

We introduce an infinite elastic solid, undisturbed for $t < 0$, and assume that a force $f(t)$ is applied at the origin of the coordinate system in the direction of the $x$ (positive) axis as shown in fig. 3.2a. Maruyama (1968), referring to Keilis-Borok (1950), derived the displacement field for $t \geq 0$ as

$$u = \frac{1}{4\pi\rho} \left[ \frac{\partial^2}{\partial x^2} (\phi - \psi) + \nabla^2 \psi \right],$$

$$v = \frac{1}{4\pi\rho} \frac{\partial^2}{\partial x \partial y} (\phi - \psi), \tag{3.1}$$

$$w = \frac{1}{4\pi\rho} \frac{\partial^2}{\partial x \partial z} (\phi - \psi),$$

where $u$, $v$, $w$ are the displacement components in the $x$, $y$, $z$ directions, respectively, and $\rho$ is the density. $\phi$ and $\psi$ represent spherical waves of the form

$$\phi = (1/r)F(t - (r/v_p)), \quad \psi = (1/r)F(t - (r/v_s)), \tag{3.2}$$

which propagate to a distance $r$ with the velocities of P- and S-waves given by $v_P$ and $v_S$, respectively. $F(t)$ is related to the force at the origin in the following manner,

$$f(t) = \mathrm{d}^2 F(t)/\mathrm{d}t^2. \tag{3.3}$$

Equation (3.1) is useful for studying various force systems at a point source. If the forces shown in figs. 3.2b, c, d and e are expressed as combinations of simple forces such as that shown in fig. 3.2a, then the displacement field can be calculated using (3.1) providing the correct substitutions are made for $x$, $y$, and $z$ (see the following example).

The single force of the type shown in fig. 3.2a is one of the simplest mathematical models of a source, although it is not very plausible from a physical point of view. The external application of a force, as supposed in this model, is unlikely to occur in a natural earthquake. It is more probable, in principle, that the force action is of a self-balancing type, that is to say, a pair of opposite forces act simultaneously on two adjacent portions in the medium, so that the resultant force is equal to zero. In fig. 3.2b, for example, a pair of forces of equal magnitude is applied to a pair of neighbouring points on either side of the origin. Let us assume that the points are located on the $x$-axis with a short distance between them, $\varepsilon$, and that the forces act in the positive and negative directions on the same axis. They balance each other and give no resultant force externally.

Displacements due to the positive force $f(t)$ acting at the point $(\varepsilon/2, 0, 0)$ may be obtained from (3.1) immediately by replacing $x$ by $x - \varepsilon/2$ (fig. 3.2b). For example, the displacement component $u$ at the point $P(x, y, z)$ is given using the expression for $u$ in (3.1) with $x$ replaced by $x - \varepsilon/2$, i.e. $u(x, y, z)$ is replaced by $u(x - \varepsilon/2, y, z)$. Similarly, we obtain $-u(x + \varepsilon/2, y, z)$ for the displacement component $u$ due to the negative force $-f(t)$ acting at the point $(-\varepsilon/2, 0, 0)$. Combination of these two components gives an expression for the double force shown in fig. 3.2b.

To study the limiting case, we introduce

$$m(t) = \varepsilon f(t), \tag{3.4}$$

and increase $f(t)$ while letting $\varepsilon$ go to zero, so that the product $m(t)$ remains finite. Following this procedure, we replace $f(t)$ in (3.3) by $m(t)$ and divide $u$ by $\varepsilon (\varepsilon \to 0)$ to satisfy the condition (3.4). That is:

$$\lim_{\varepsilon \to 0} \{[u(x - \varepsilon/2, y, z)/\varepsilon] - [u(x + \varepsilon/2, y, z)/\varepsilon]\} = -\partial u(x, y, z)/\partial x. \tag{3.5}$$

Other displacement components may be obtained in a similar manner, and we conclude that the displacements due to the force couple shown in fig. 3.2b are $(-\partial u/\partial x, -\partial v/\partial x, -\partial w/\partial x)$, i.e. they are obtained by differentiation of the displacements given by (3.1) due to the single force.

Another type of force couple, shown in fig. 3.2c, can be treated in a similar manner. In this case, $(-\partial u/\partial y, -\partial v/\partial y, -\partial w/\partial y)$ is the displacement field due to the force couple. For simplicity, we change notation so that the torque of a force couple is denoted by $f(t)$ and the displacement due to it by $(u, v, w)$. Then,

$$u = -\frac{1}{4\pi\rho} \left[ \frac{\partial^3}{\partial x^2 \partial y} (\phi - \psi) + \frac{\partial}{\partial y} \nabla^2 \psi \right],$$

$$v = -\frac{1}{4\pi\rho} \frac{\partial^3}{\partial x \partial y^2} (\phi - \psi), \tag{3.6}$$

$$w = -\frac{1}{4\pi\rho} \frac{\partial^3}{\partial x \partial y \partial z} (\phi - \psi).$$

Forces in the present system balance one another, as in the previous case. This system, however, has a torque about the $z$-axis. This type of force system is called a *single-couple* to distinguish it from the type discussed next.

Further application of the theory given above enables us to analyse a more complicated system such as that shown in fig. 3.2e, which is called a *double-couple* type. As shown in the figure, this type is a combination of two single-couple systems whose torques balance. Therefore, its displacement field is derived as

$$u = -\frac{1}{4\pi\rho}\left[2\frac{\partial^3}{\partial x^2 \partial y}(\phi - \psi) + \frac{\partial}{\partial y}\nabla^2\psi\right],$$

$$v = -\frac{1}{4\pi\rho}\left[2\frac{\partial^3}{\partial x \partial y^2}(\phi - \psi) + \frac{\partial}{\partial x}\nabla^2\psi\right], \qquad (3.7)$$

$$w = -\frac{1}{4\pi\rho}\left[2\frac{\partial^3}{\partial x \partial y \partial z}(\phi - \psi)\right].$$

Displacements for the model shown in fig. 3.2*d* are also expressed by (3.7), proving the equivalence of the models shown in fig. 3.2*d* and *e* in their mechanical effect. Note that the force axes in fig. 3.2*d* are rotated about the *z*-axis by 45° relative to those in fig. 3.2*b*.

We learn from the analysis above that various types of source force systems may be interpreted on the basis of a single force system. These types of force system at a point source are generally called *nuclei of strain*, and are used to construct advanced models of earthquakes.

### 3.1.3   Radiation patterns

Let us follow Maruyama (1968) and continue to study wave radiation from a double-couple source. To evaluate the right-hand side of (3.7), we only retain terms of the order $(1/r)$ after differentiating and neglect higher-order terms (proportional to $1/r^2$ and $1/r^3$) as they contribute little at a great distance. This makes the mathematics extremely simple (for example, $\partial\phi/\partial x$ in the equations may be approximated by $(-1/v_{\mathrm{p}})(x/r^2)F'$ because the derivative of $\phi$ $(=F/r)$ with respect to $x$ may be replaced by $(1/r)\partial F/\partial x$, and the term in $\partial(1/r)/\partial x$ may be neglected).

Thus, for a P-wave, we obtain from (3.7)

$$u = \frac{1}{4\pi\rho}\frac{1}{v_{\mathrm{P}}^3}\frac{2x^2 y}{r^3}\frac{1}{r}f'\left(t - \frac{r}{v_{\mathrm{p}}}\right),$$

$$v = \frac{1}{4\pi\rho}\frac{1}{v_{\mathrm{P}}^3}\frac{2xy^2}{r^3}\frac{1}{r}f'\left(t - \frac{r}{v_{\mathrm{P}}}\right), \qquad (3.8)$$

$$w = \frac{1}{4\pi\rho}\frac{1}{v_{\mathrm{P}}^3}\frac{2xyz}{r^3}\frac{1}{r}f'\left(t - \frac{r}{v_{\mathrm{P}}}\right),$$

and, for an S-wave

$$u = \frac{1}{4\pi\rho} \frac{1}{v_S^3} \frac{y(-x^2+y^2+z^2)}{r^3} \frac{1}{r} f'\left(t - \frac{r}{v_S}\right),$$

$$v = \frac{1}{4\pi\rho} \frac{1}{v_S^3} \frac{x(x^2-y^2+z^2)}{r^3} \frac{1}{r} f'\left(t - \frac{r}{v_S}\right),$$  (3.9)

$$w = \frac{1}{4\pi\rho} \frac{1}{v_S^3} \frac{(-2xyz)}{r^3} \frac{1}{r} f'\left(t - \frac{r}{v_S}\right),$$

where, from (3.3), $f'(t - r/v_P)$ replaces $F'''(t - r/v_P)$.

In order to transform the Cartesian coordinates $(x, y, z)$ in (3.8) and (3.9) into the polar coordinates $(r, \theta, \phi)$, see fig. 3.3, we take

$$x = r \sin\theta \cos\phi,$$
$$y = r \sin\theta \sin\phi,$$  (3.10)
$$z = r \cos\theta.$$

Then, $x, y, z$ are differentiated with respect to $r, \theta, \phi$ and the direction cosines of the $r$-, $\theta$-, and $\phi$-directions in the Cartesian coordinates $(x, y, z)$ are obtained as $(\sin\theta\cos\phi, \sin\theta\sin\phi, \cos\theta)$, $(\cos\theta\cos\phi, \cos\theta\sin\phi, -\sin\theta)$ and $(-\sin\phi, \cos\phi, 0)$, respectively (each quantity increasing in the positive direction). Therefore, the displacement components in the $r$-, $\theta$-, $\phi$-directions may be written as

$$u_r = u \sin\theta \cos\phi + v \sin\theta \sin\phi + w \cos\theta,$$
$$u_\theta = u \cos\theta \cos\phi + v \cos\theta \sin\phi - w \sin\theta,$$  (3.11)
$$u_\phi = -u \sin\phi + v \cos\phi.$$

Substituting (3.8) and (3.9) in (3.11) and using (3.10), we finally obtain

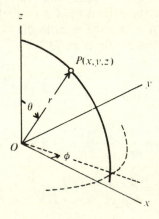

Fig. 3.3. Cartesian and polar coordinates.

$$u_r = \frac{1}{4\pi\rho} \frac{1}{v_P^3} \frac{1}{r} f'\left(t - \frac{r}{v_P}\right) \sin^2 \theta \sin 2\phi,$$

$$u_\theta = \frac{1}{4\pi\rho} \frac{1}{v_S^3} \frac{1}{r} f'\left(t - \frac{r}{v_S}\right) \sin \theta \cos \theta \sin 2\phi, \qquad (3.12)$$

$$u_\phi = \frac{1}{4\pi\rho} \frac{1}{v_S^3} \frac{1}{r} f'\left(t - \frac{r}{v_S}\right) \sin \theta \cos 2\phi,$$

as the wave radiation from a double-couple source. This is practically equivalent to expressions derived in earlier work (see for example Honda, 1957).

We remark first that $u_r$ is responsible for the P-wave whereas $u_\theta$ and $u_\phi$ produce the S-wave. Although these two different disturbances are propagated with different velocities, $v_P$ and $v_S$, their wave-forms given by $f'(t)$ will be the same. As previously, $f(t)$ represents a time function of the torque. Therefore, we shall observe displacements in a wave-form proportional to the time derivative of the torque at the source (see fig. 3.4.).

The azimuthal distribution of the amplitude of the displacement should also be noted. Patterns of the displacement distribution are shown in fig. 3.5. Figs. 3.5a and b represent the patterns of $u_r$ (P-wave) and of $u_\theta$ and $u_\phi$ (S-wave), respectively. Let $\theta = \pi/2$ in (3.12) so that the distribution of $u_r$ and $u_\phi$ in the equatorial plane may be studied ($u_\theta = 0$, in the present case). Suppose $f'(t - r/v_P)$ is positive, $u_r$ becomes positive for the azimuths, $0 < \phi < \pi/2$ and $\pi < \phi < 3\pi/2$, but negative for $\pi/2 < \phi < \pi$ and $3\pi/2 < \phi < 2\pi$. Thus, the pattern of $u_r(\phi)$ is of the quadrant type with nodal lines at $\phi = 0$, $\pi/2$, $\pi$ and $3\pi/2$, respectively. The pattern of $u_\phi$ is similar to that of $u_r$, but its nodal lines appear at $\phi = \pi/4$, $3\pi/4$, $5\pi/4$ and $7\pi/4$. The amplitudes of $u_r$ and $u_\phi$ change with $\theta$ as functions of $\sin^2\theta$ and $\sin \theta$, respectively, but their nodal points are the same for $0 < \theta < \pi$. In a three-dimensional view, therefore, the radiation pattern of the P-wave is characterized by a pair of orthogonal *nodal planes* which separate

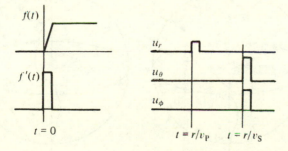

Fig. 3.4. Wave-forms of seismic signals (on a relative scale), generated from a double-couple source with a ramp time function.

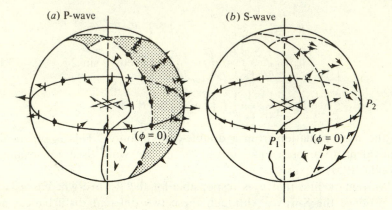

Fig. 3.5. Wave radiation displacements from a double-couple source.

compressions from dilations (fig. 3.5). The $u_\phi$ component shows a similar pattern, but the $u_\theta$ component does not (see (3.12)). As a result, the radiation pattern of the S-wave looks more complex than that of the P-wave.

The results which were obtained from (3.7) are for a double-couple point force system. If we now consider a single-couple type, for historical interest (§3.3.1), we may use (3.6) and express the wave radiation as follows:

$$u_r = \frac{1}{4\pi\rho}\,\frac{1}{v_P^3}\,\frac{1}{r}\,f'\!\left(t - \frac{r}{v_P}\right)\frac{\sin^2\theta\,\sin 2\phi}{2},$$

$$u_\theta = \frac{1}{4\pi\rho}\,\frac{1}{v_S^3}\,\frac{1}{r}\,f'\!\left(t - \frac{r}{v_S}\right)\frac{\sin\theta\,\cos\theta\,\sin 2\phi}{2}, \qquad (3.13)$$

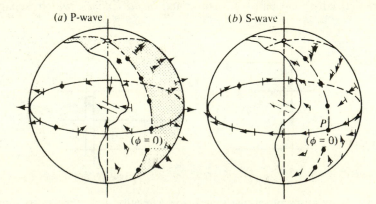

Fig. 3.6. Wave radiation displacements from a single-couple source.

$$u_\phi = -\frac{1}{4\pi\rho}\frac{1}{v_s^3}\frac{1}{r}f'\left(t-\frac{r}{v_s}\right)\sin\theta\sin^2\phi.$$

Comparing this with (3.12), we see that the single-couple type is effectively equivalent to the double-couple type with respect to $u_r$ and $u_\theta$. However, the single-couple $u_\phi$ component is characterized by the $\sin^2\phi$ factor in the azimuthal distribution in contrast to the $\cos 2\phi$ factor in the previous model. As fig. 3.6 illustrates, S-waves from the single-couple source have nodes at $\phi = 0$ and $\pi$ but do not change polarity across them.

### 3.2 FAULT-PLANE SOLUTION

#### 3.2.1 *Elementary considerations*

Separation of compressions from dilatations on a map is relatively easy in cases as that shown in fig. 3.1, when many stations are located around the source. If there are no stations around the source, which is often the case, we must use remote stations. For this purpose, Byerly (1955) proposed a graphical technique to use data from a world-wide distribution of stations. Instead of a pattern analysis of a local map as in fig. 3.1, his method projects a global distribution of data onto a plane for comparison with theoretical nodal lines (refer to the review papers listed in §3.1.1 for details). Other projection techniques have since been proposed, and the most important of these are discussed below.

Let us surround the focus with a unit sphere, which is sometimes called the *focal sphere* (fig. 3.7). A seismic ray reaching a station $S$ leaves the sphere with an angle of emergence $i$, which is a function of the epicentral distance $\Delta$. We assign $S'$ on the sphere with the polarity of the initial P-phase as observed at $S$, then $S'$ represents the data at $S$. Repeating this procedure for all the station data available for a seismic event, we obtain a comprehensive view of wave radiation from the focus. The initial character of an elastic wave should, in principle, remain unchanged during its propagation, so the evidence on the focal sphere should faithfully represent the original polarity of wave radiation. Suppose the source mechanism is one of those discussed in §3.1.3, the P-wave polarity on the focal sphere should then be separated by a pair of orthogonal planes into four regions of alternate signs, compressional and dilatational.

In practice, most fault-plane work involves a procedure of geometrical projection so that we may display the focal sphere and observational data on a two-dimensional diagram. Several techniques are available for projection work. The *stereographic projection*, which is sometimes called the *Wulff net*, takes a conformal azimuthal projection. It projects the arc

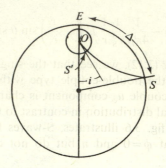

Fig. 3.7. A seismic ray in the Earth (*OS*) and a focal sphere (small circle). *O* is the focus, *E* is the epicentre, and *S'* represents the projection of *S* on the surface of the focal sphere. Note that a seismic ray is drawn as concave (rather than as a straight line) because the velocity of seismic waves generally increases with depth in the Earth.

of a great circle (e.g. the intersection of the focal sphere by a nodal plane) into an arc on the diagram (see Stauder, 1962). Another popular scheme is the *equal-area projection* (*Schmidt net*), which has the advantage that relative areas on the focal sphere may be determined easily (Honda, 1957, see below).

### 3.2.2  *Graphical techniques*

Before studying practical techniques, we must consider basic rules for mapping the nodal planes. First, a nodal plane intersects the focal sphere so that a nodal line is a great circle on the projection and therefore it must lie on a meridian of the projection net. Second, if one of the nodal planes is given (for example, plane *A* in fig. 3.8*a*), another plane, *B*, has constraints imposed on it by its orthogonal relation to plane *A*; in other words, plane *B* must contain the pole (*A*) of the first plane, and vice versa, and must reasonably explain the observed polarities. The figures 5(*a*)–(*f*) in the following describe in detail the steps required to impose this constraint.

After gathering the initial motion data from many stations all over the world, the procedures given below are followed to arrive at a solution (Ritsema, 1959; Japan Meteorological Agency, 1971). Polarity of the initial P-wave is the basic datum for this work, but other reflected waves, such as pP and PP, may also be used after proper correction for propagation effects (e.g. reversal of polarity by reflection at a free surface).

1. Determine the azimuth, $\phi_0$, and distance, $\Delta$, of a station from the epicentre.

2. Use $\Delta$ and determine the angle of emergence, $i$, from the focus for

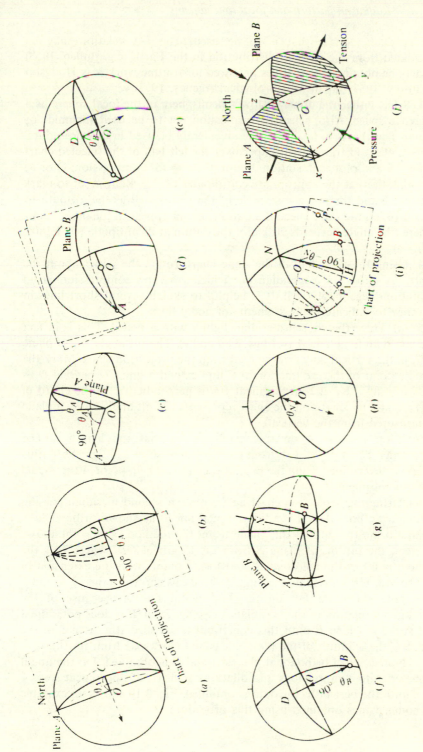

Fig. 3.8. The procedure for mapping a fault plane diagram.

the particular phase that is to be used. The $i$–$\Delta$ relation may be calculated from the velocity distribution in the Earth (e.g. Bullen, 1953) or may be obtained from tables prepared for routine work (e.g. Hodgson & Storey, 1953; Japan Meteorological Agency, 1971).

3. Select either the upper or lower hemisphere of the focal sphere as a reference space. Also, select the projection net to be used (Schmidt or Wulff). The charts in fig. 3.9 may conveniently be used in this procedure. Place a sheet of transparent paper over the left half of the selected chart in fig. 3.9 and plot each station datum (C or ● for compression; D or ○ for dilatation) at the appropriate coordinates $(\phi_0, i)$. Remember to mark the centre and the outermost circle of the net as well as the azimuth to the north, on the same sheet. If a datum should appear in the other hemisphere (the angle $i$ exceeds 90°), plot the datum at its antipode $(\phi_0 + 180°, 180° - i)$.

4. Using the right half of the same chart, rotate the paper about the centre to search for a meridian line which separates compressions from dilatations reasonably well. It is helpful to sketch several short lines in advance to indicate likely segments of nodal lines.

5. (a) Trace the chosen meridian lightly with a pencil. This is in fact the meridian for plane $A$ (see figs. 3.8a and c). The tilt angle, $\theta_A$, of plane $A$ from the vertical may now be read from the projection chart. Mark the middle point of the arc (meridian) $C$ and extend a line $CO$ beyond $O$ as shown in fig. 3.8a (the line segment $CO$ is perpendicular to the chord of this arc and its length on the chart represents the tilt angle of $\theta_A$ of plane $A$ measured from the vertical).

(b) Plot a point $A$ on this extension so that the length of $OA$ represents the angle $90° - \theta_A$ from the vertical as shown in fig. 3.8b, c (this length is determined using the projection chart). Then, point $A$ represents the pole of plane $A$.

(c) Using the projection chart again, draw a second meridian passing through the pole $A$ as a candidate nodal line $B$ as shown in fig. 3.8d.

(d) Locate the pole of this plane using the method of step (b) above; i.e. read the tilt, $\theta_B$, of plane $B$ from the length of $OD$ (where $D$ is the middle point of the meridian $B$) and locate point $B$ on the extension of $DO$ so that $OB$ represents the angle $90° - \theta_B$, as shown in figs. 3.8e and f. The pole $B$ should fall on the first nodal line $A$, because of the orthogonal relation of the two planes (see fig. 3.8g). If it does not, adjust the strike of plane $B$ until this constraint is satisfied (the 'strike' of the plane is the azimuth of the plane measured clockwise from north).

6. Now a set of orthogonal planes $A$ and $B$ is obtained. Do the nodal lines separate compressions and dilatations easily? If not, repeat steps 4 to 6 until the best fit to the data is obtained. Fig. 3.10 shows an example of nodal planes obtained using this procedure.

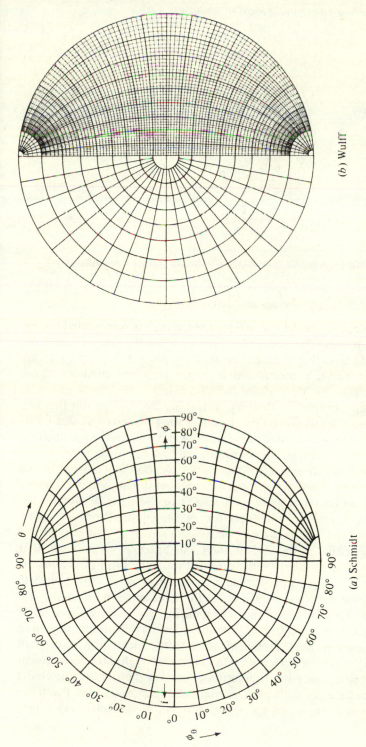

(a) Schmidt

(b) Wulff

Fig. 3.9. (a) The equal-area projection (Schmidt net). (b) The stereographic projection (Wulff net). The left-hand halves of the projections illustrate the azimuth $\phi_0$, and the angle of emergence, $i$, to a station, and are used for plotting station data (see step 3 in §3.2.2). The right-hand halves give families of equi-latitude ($i$) lines and of meridians (equi-$\phi$ lines). A nodal line, which makes a great circle on the focal sphere, appears on the diagram as a meridian, so that the tilt of the fault plane from the vertical may be read from the scale given along the equator of the right halves, $\phi$ to a point on a meridian may also be read using this diagram by referring to the equi-latitude lines.

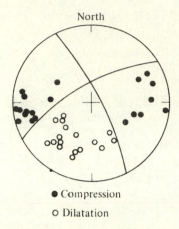

Fig. 3.10. Example of a fault-plane solution (upper hemisphere).

### 3.2.3  *Basic parameters and interpretation*

If the double-couple model (fig. 3.2*e*) is used, a fault is represented by one of the nodal planes. The nodal plane representing the fault is called the *fault plane* and the other is called the *auxiliary plane*. Let us return to fig. 3.8*g*, and assume the planes *A* and *B* are the fault and auxiliary planes respectively. The line $ON$, where the two planes cross one another, is called the null axis ($\overrightarrow{ON}$ is the null vector), as the medium on this line experiences no displacement. Orientation of each plane is specified by its strike and dip angle, whereas orientation of the null axis is specified by its azimuth and inclination. This geometrical system has three degrees of freedom about the three axes of the Cartesian coordinates fixed to the sphere. Therefore, it is fixed by three independent orientation parameters. If one more parameter is given with respect to the sense of motion, the pattern of compression and dilatation of the initial P-phase is fully specified. Thus, we obtain a *fault-plane solution*.

As an alternative to the system shown in fig. 3.2*e*, we may use that shown in fig. 3.2*d* to interpret the observations. The focal mechanism is described, in this case, by pressure and tension axes. As shown in fig. 3.8*j*, these axes lie on the planes which bisect the solid angles between the two nodal planes. To solve the problem graphically, we first mark a point *H* on the extension of *NO* (fig. 3.8*i*) so that the length *OH* on the projection plane represents an angle complementary to the inclination of the null axis (i.e. *ON*). Next, we draw a meridional line through *AHB* to represent the *xy*-plane and, using the equi-latitude lines on the right of the charts in fig. 3.9, locate a point $P_1$ on it so that the angles $AOP_1$ and $P_1OB$ are 45°. The point $P_2$ is set on the same meridian in a similar way. Then,

$P_1O$ and $P_2O$ indicate the directions of the compressional and dilatational forces respectively. Special care is taken to assign the force with
a sense consistent with the wave polarity (tension to the compression
lune; pressure to the dilatation lune), otherwise confusion may occur in
interpretation of the data.

A fault-plane solution helps us to learn a great deal about the physical
and tectonic conditions in the source region (see later chapters). A brief
account of the interpretation of solutions follows for future reference.
Fig. 3.11 demonstrates typical patterns of solutions (viewed from above)
for three elementary cases (which are interpreted, in terms of the fault's
behaviour, in §4.1.1), (*a*) *strike-slip* faulting in a vertical plane, and (*b*, *c*)
*dip-slip* faulting in a plane dipping at 45°. If the dip-slip is the *reverse*
type, compressional polarity is observed about the centre of the diagram
(fig. 3.11*b*); if it is the *normal* type, dilatation polarity is observed (fig.
3.11*c*). A fault-plane solution tells us about two possible planes of
faulting, but does not specify a unique one. In order to find the plane of
faulting, therefore, some supplemental information such as geological
trends, linear aspects of aftershock distribution, etc., must be given. If the
preferred plane of faulting is found to be plane *A* (see fig. 3.11*a*) the
polarity is said to be in the *left-lateral* sense; if *B* is the preferred plane,
the polarity is said to be in the *right-lateral* sense. A more general
solution type is a mixture of (*a*) with (*b*), or (*a*) with (*c*), as shown in fig.
3.10. Even in these cases, the sense of dip-slip (reverse or normal) is easily
found by examining which polarity (compression or dilatation) occupies
the central area of the fault-plane solution. The direction of the fault slip
is known from geometrical considerations to be parallel to *AO* or *BO* in
fig. 3.8*g* (note that this is different from the direction of the pressure axis
which can be seen in fig. 3.8*j*). If the slip amplitude is known, we may
assign this direction to it to define a *slip vector*.

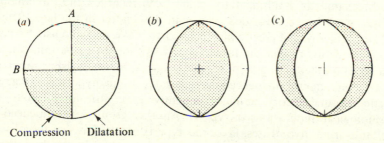

Fig. 3.11. Elementary types of fault-plane solution: (*a*) strike-slip; (*b*) dip-slip
(reverse); and (*c*) dip-slip (normal).

### 3.2.4    *Limitations of the technique*

Several fundamental difficulties that limit this technique must be mentioned. The first is rather technical and attributed to the quality of observational data. A fault-plane solution cannot, in general, be free from contamination with incorrect readings of the polarity. Incorrect polarity readings can arise at a number of stations and are due mainly to low signal-to-noise ratios, but are sometimes due to the accidental reversal of the instrumental polarity. These data make precise determination of the nodal lines difficult, especially when they appear close to the lines on the projection. Recent developments in the global seismographic network have improved the data quality significantly, yet this difficulty remains in analyses of smaller earthquakes. Processing of inconsistent readings is a serious problem in computerized techniques (§3.4).

As noted in the foregoing discussion, it is impossible to discriminate between the single- and double-couple models by the P-wave solution alone. And, even if we accept the fault-origin model, a solution for the P-wave does not indicate which plane represents the fault. These problems are due to the fact that the force systems of single- and double-couple types are equivalent with respect to P-wave radiation. Further data must be provided to discriminate between these force systems.

### 3.3    SINGLE- AND DOUBLE-COUPLE MODELS

### 3.3.1    *Conflicting views in focal mechanism studies (historical sketch)*

The previous discussion has introduced two conflicting views of the force system at a focus. To distinguish them, we sometimes use the terms *type I* and *type II*, after Honda (1957), denoting a single-couple and a pair of perpendicular coplanar couples, respectively. The latter type is equivalent to a double-couple type in its seismic effect (fig. 3.2). A Symposium on the Mechanics of Faulting (Hodgson, 1959) recorded the fundamental disagreement between various groups in this matter in the late 1950s.

The conflict was of great concern to earthquake seismologists, because a preference for one of the force types was thought to be immediately related to a difference in view on earthquakes. Seismologists supporting type I were mainly influenced by the fault-origin hypotheses, e.g. the elastic rebound theory, and consequently, gave little attention to the seismological implications of type II. Another group, less influenced by the fault-origin hypotheses, favoured type II on the basis of the observations, and as a consequence, were inclined to doubt the hypotheses which were supported by the first group.

With the benefit of hindsight, we can see that the problem would have been most quickly resolved by the consideration of two separate questions: first, which type best explains the radiation pattern of the S-wave; and second, can all earthquakes be attributed to faulting. The dispute between the two groups has ended in a draw. As to the first question, the superiority of type II has been proved clearly (see §3.3.2 and §3.3.3). As to the second question, however, fault-origin hypotheses such as the elastic rebound theory have been basically accepted, at least for big shallow earthquakes (§6.1.2). Contrary to the earlier beliefs of seismologists, these two conclusions are not in conflict with one another, because the force–dislocation equivalence has proved that land movement, such as a fault offset, is well represented by the force type II, but not type I (§4.3.2).

Further discussion of this problem will appear in §§3.3.2, 3.3.3, 4.3.2 and 6.1.2. Regarding the first question, a radiation pattern for S is the key to a uniquely-defined solution, as noted on several occasions in the discussion above. In fact, the symposium mentioned above (Hodgson, 1959) stressed the need for agreement among seismologists on the interpretation of S. Nevertheless, it took several more years before agreement was reached. Let us now study a few examples of advanced techniques used to recognize type II forces from observed data.

### 3.3.2 *S-wave and its polarization angle*

The extended confusion in the study of the S radiation pattern may be undoubtedly attributed to the difficulty in identification of the initial S-phase. As seismologists know, disturbances following the arrival of P and related phases at the measuring device often obscure the onset of S causing errors in the polarity readings. Various techniques have been introduced in an attempt to reduce the risk of this type of error. For example, techniques based on sign combination for P, SV, SH, etc. (Keilis-Borok, 1959), amplitude ratio of P to S (Honda, 1957), and synthesis of pulse seismograms (Honda, 1957; Kasahara, 1963) have been suggested.

Another technique uses the *polarization angle*, $\varepsilon$ of the S-wave, defined as

$$\varepsilon = \tan^{-1} (u_H/u_V), \tag{3.14}$$

where $u_H$ and $u_V$ are the SH and SV components of the S motion, respectively. In other words, $\varepsilon$ is the angle that the direction of S motion makes with the vertical plane of the incident seismic radiation (fig. 3.12). To derive $\varepsilon$ from a set of seismograms, the recorded amplitudes must be corrected for the free surface conditions (see for example Bullen, 1953). Denoting the components of the incident S-wave, along and perpendicular to the great circle path at the station as $\bar{u}_R$ and $\bar{u}_H$,

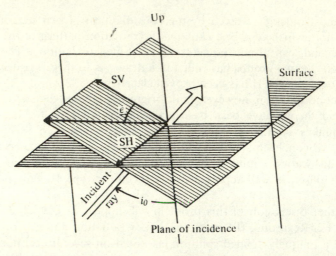

Fig. 3.12. The polarization angle, $\varepsilon$, of the S-wave. The seismic vibration of the medium due to the S-wave has two components, SV and SH, parallel and transverse to the seismic ray respectively. The SV component is in the vertical plane of incidence, and the SH component is characterized by the horizontal movement of the medium. (From Stauder, 1962.)

respectively, we may write

$$\gamma_0 = \tan^{-1}(\bar{u}_H/\bar{u}_R) \tag{3.15}$$

where $\gamma_0$ is the angle which is usually measured and assumed to be equal to the polarization angle, $\varepsilon$. In fact, the two angles are not equal, but are related by the equation

$$\tan \gamma_0 = \left\{ \frac{\cos^2 2j_0 + 2 \sin j_0 \sin (2j_0) [(\beta_0/\alpha_0)^2 - \sin^2 j_0]^{1/2}}{\cos j_0 \cos^2 2j_0 + \sin j_0 \sin 2j_0 \cos 2j_0} \right\} \tan \varepsilon$$
$$= f(j_0, \beta_0/\alpha_0) \tan \varepsilon, \tag{3.16}$$

where $\alpha_0$ and $\beta_0$ are the P- and S-wave velocities at the surface, and $j_0$ is the angle of incidence of the wave (Nuttli & Whitmore, 1962). If $j_0$ is greater than $\sin^{-1}(\beta_0/\alpha_0)$, total reflection occurs and $f(j_0, \beta_0/\alpha_0)$ becomes complex. To avoid this, polarization angles are studied only within a certain range of epicentral distance. It is fortunate that $f$ is approximately equal to 1 under the usual observational conditions. Nuttli & Whitmore calculated $f$ for several typical sets of the parameters and recommended that the epicentral distance should be greater than 45 from the focal point of the source when using the technique described above.

Theoretical distributions of the polarization angles of S in a focal sphere can be studied from the force systems discussed in §3.1 where characteristic patterns are obtained for the single- and double-couple models.

This allows us to discriminate between the two models when comparing data compiled by the polarization angle technique. Fig. 3.13 demonstrates a successful example of the use of this technique by Udias (1964) on an earthquake off the coast of Kamchatka (22 July 1953). Notice the good agreement of observations with theory, when a proper fault-plane solution has been found. Actually, the direction of the short bars in the diagram gives the polarization angle with azimuth to the epicentre from the station. A family of thin curved lines represents the theoretical direction of the polarization, connecting the projections of the pressure and tension axes of the fault-plane solution. The fault-plane solution is illustrated by a pair of nodal line arcs (thick arc lines) on the stereographic projection in fig. 3.13.

This technique is based on the amplitude ratio of the SH and SV

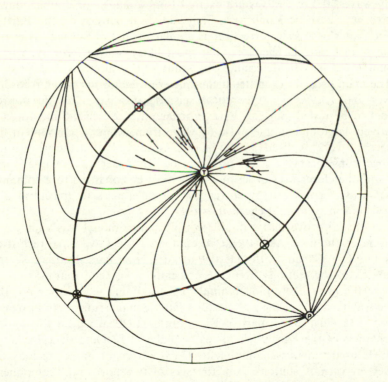

Fig. 3.13. Numerically-derived solution to the fault-plane model for an earthquake off the coast of Kamchatka (22 July 1953). A double-couple source has been assumed. The short line segments represent observed polarizations. The continuous thin lines represent the theoretical polarization directions for the fault-plane solution indicated by a pair of nodal line arcs (thick lines) and the axes of tension T and pressure P. (From Udias, 1964.)

components, which is less sensitive to the accuracy of polarity reading (but phase correlation between the two components must hold). Another advantage of this technique is its adaptability to computer methods. The least-squares method and other numerical techniques for finding an optimum solution may be applied more easily to continuous data like the polarization angle than to discrete data like the polarity of initial motions. This is especially advantageous when computers are used to determine the focal mechanism (§3.4).

### 3.3.3 *The use of surface waves*

The use of surface waves, which has been made possible by recent progress in the theory and observation of surface wave excitation, provides a most straightforward approach to the study of seismic source problems. Radiation of surface waves is closely related to that of seismic body waves. If Love waves are excited from a single- or double-couple source with a vertical null axis, their radiation pattern on the Earth's surface, after necessary correction, is similar to that of S-waves from the same source model. Therefore, the most likely source force type in a seismic event may be determined by study of the surface waves.

The main procedures in this technique are *equalization* of the recorded waves and *synthesis* of theoretical seismograms based on the source model to be tested. Surface waves appear on a teleseismic record as a relatively large and simple pulse or wave train, which is convenient for numerical work. A disadvantage of this, however, is that the waveforms change considerably during propagation. In order to make a comparison, therefore, a set of seismograms must be equalized to a common observational condition by correction for differences in epicentral distances and instrumental characteristics.

In order to prove that a double-couple source model was valid for a Nevada earthquake, Aki (1960*a*, *b*) compared the Love-wave record at Resolute Bay, Canada with that at Palisades, New York. In this case, the azimuths of the stations from the epicentre were such that Aki could choose the most likely model simply by examining whether or not the Love-wave records at the two stations (after equalization) were in phase.

Fig. 3.14 illustrates a part of Kanamori's (1970) work on the source mechanism of the Kurile Islands earthquake of 13 October 1963 ($M = 8.3$). He used transverse components (Love waves, or more precisely $G_4$ waves) at various stations from the sets of two horizontal component records. The records were equalized to a propagation distance of $7\pi/2$ (where the angle is measured about the Earth's centre) and to a trace amplitude on the standard long-period seismogram with magnification

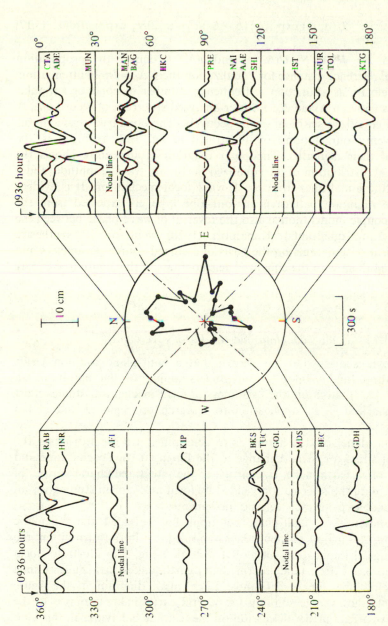

Fig. 3.14. Radiation of Love waves at various azimuths from the Kurile Islands earthquake (13 October 1963, $M = 8.3$). Trace amplitudes of Love waves ($G_4$) at the WWSSN stations (the station names and locations are given in appendix 3) are compared after equalization of propagation distance and instrumental magnification. The circular diagram at the centre of the figure is the polar plot of the maximum trace amplitude versus azimuth to the stations. (From Kanamori, 1970.)

of 1500, by the following formula:

$$u(\Delta_0, t) = (|\sin \Delta/\sin \Delta_0|)^{1/2} \exp [K(\Delta - \Delta_0)]$$

$$\times \int_{-\infty}^{\infty} U(\Delta, \omega) \exp i\{\omega[(\Delta - \Delta_0)/c] - (\pi/2)N\} \exp (i\omega t)d\omega \quad (3.17)$$

where $\Delta$, $\Delta_0$, $K$, $c$, $N$, $U(\Delta, \omega)$ and $u(\Delta_0, t)$ denote the propagation distance, reference distance for equalization (here $\Delta_0 = 7\pi/2$), attenuation, phase velocity as a function of frequency, number of polar or antipolar passages between $\Delta$ and $\Delta_0$, Fourier-analysed spectrum of record $u(\Delta, t)$, and the equalized seismogram respectively. The equalized seismograms from twenty-four stations are illustrated in fig. 3.14, arranged in the order of their azimuths. The polar diagram at its centre shows the azimuthal distribution of the equalized maximum trace amplitude, and a pair of the nodal lines from the P-wave focal mechanism. It is evident that the radiation pattern is of four-lobe type, as expected from the double-couple model (notice also the reversal of wave phase between the neighbouring quadrants). Kanamori (1970) went on to synthesize theoretical seismograms on the basis of assumed source parameters and compared them with the equalized seismograms in this figure (see §5.5.4).

### 3.4   APPLICATIONS OF FAULT-PLANE SOLUTIONS

#### 3.4.1   *Fault-plane data and the Earth's tectonics*

The previous discussion has shown that a fault-plane solution yields information fundamental to our understanding of the nature of an earthquake. In addition, the usefulness of focal mechanism studies must be emphasized in its application to research on regional and global tectonics.

A fault-plane solution allows us to determine, at a great distance, the mode of stress (or, strain) change in the Earth, at the specific time and locality of an earthquake. In contrast to the integrated characteristics of geodetic data the information yielded by focal mechanism study gives an instantaneous picture of tectonic movements.

Accumulation of fault-plane solutions has revealed the kinematic behaviour of major seismic zones, and led us into research on the seismological implications of global tectonics. Fig. 3.15 illustrates focal mechanisms of deep (and intermediate) earthquakes in a huge seismic zone below the Japanese Islands. Each small diagram in the figure represents the average solution for several earthquakes grouped in the respective area. Further discussion on the tectonic behaviour in this kind of seismic zone will appear in chapter 7.

Sykes (1967), in a comparative study of focal mechanisms of many shallow earthquakes along the mid-Atlantic ridge, sought to test the

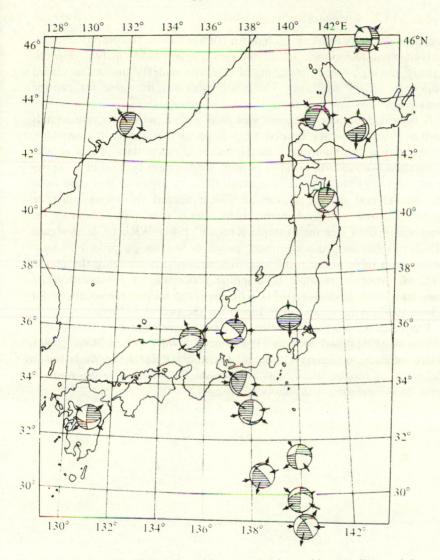

Fig. 3.15. Geographic distribution of focal mechanisms of intermediate and deep earthquakes in and near Japan. Each small diagram shows the upper hemisphere of the average solution for several earthquakes grouped in that area. Heavy arrows represent the axes of tension and pressure. (From Ichikawa, 1971.)

transform fault model hypothesized by Wilson (1965). After careful location of the focal points of earthquakes along the ridge, Sykes confirmed that the seismic events along a line segment connecting the adjoining crests of the ridge indicate the operation of a strike-slip type mechanism, providing strong support for Wilson's transform fault model (see §7.1.3).

### 3.4.2    *Computerized techniques*

Increasing activity in focal mechanism studies has led to computerization of fault-plane techniques. Its advantages are significant in two respects, namely, the high-speed processing of extensive material and an increased objectivity in the solutions. These are especially desirable for routine work.

A typical computer program will deal with a set of discrete signals such as P-wave polarity, or a continuous set such as S polarization angle, or the two in combination. Seismologically, P-wave data are of greater value than S-wave data because of their greater density, ease of measurement, and reliability. A mathematical difficulty arises in this case from the discreteness of the P signals. Some special technique must be provided, using statistical methods, to obtain a 'best fit' model to the observed P data (see for example Knopoff, 1961). Wickens & Hodgson (1967) have developed a computer program for this purpose, and have published a table of the results for 618 earthquakes covering the period 1922–62. Another example of machine processing of P-wave polarity may be seen in Ichikawa (1971), who gives the fault-plane solutions for about 1000 events in and near Japan for the period 1926–68.

Programs for S-wave solutions have been developed by Udias (1964), Hirasawa (1966), and Stevens (1967) using S-wave polarization angles. More recently, Dillinger, Pope & Harding (1971) have described a program for routine work, which processes P- and S-wave data on the basis of the maximum-likelihood method.

# 4 Earthquake faults

Fault movement associated with a big earthquake is perhaps the most spectacular evidence for living tectonics in the Earth. The activity of the San Andreas fault in 1906, for example, caused a land offset over a length of more than 400 km in California (see Iacopi, 1976, for photographs of the Californian faults).

The association of faulting with a big earthquake is not always immediately confirmed. In the case of the Alaskan earthquake of 1964, for example, no appreciable fault breakages were seen (except those which appeared in the Patton Bay area) in spite of its large magnitude. However, the land uplift and subsidence, observed extensively along the coast of the Gulf of Alaska, could be attributed to a principal fault off the coast of Alaska. If the sea-bottom deformation southeast of the Alaskan coast were traced, a giant submarine fault extending about 600 km along the Aleutian trench axis would certainly be discovered.

Fault breakage may occur even in minor earthquakes, but, in general, the probability of observing it at the ground surface decreases with decreasing magnitude (see table 4.2). A close association between faults and earthquakes has long been recognized, although the nature of this association has not been well understood until recently. The purpose of this chapter is to provide the reader with a review of this problem from the fault study side. We shall first study the general features of earthquake faults by considering their surface evidence, such as mode of offset, empirical relations of fault dimensions versus earthquake magnitude, and geodetic aspects of the displacement fields around them (§4.1). Then in §4.2, two-dimensional models will be constructed to interpret the basic fault parameters. Three-dimensional models will be studied in §4.3, enabling us to deal with more plausible fault models. To explain their mathematical foundations, an introductory discussion on the elasticity theory of dislocation is given in the first part of the section. Rockmechanical aspects of faulting are discussed in chapter 6.

### 4.1.1   *Fault behaviour*

A fault is a slip surface in the Earth, across which discontinuous land movement (or offset) takes place. Its configuration under the ground cannot be observed so we must construct a model of the fault on the basis of the little surface evidence available (e.g. the length of the fault, and its offset during an earthquake). A rectangular shape for the fault plane, with one pair of its sides parallel to the free surface, is often assumed. This assumption may be accepted as a first approximation, because the surface breakage is usually straight. In fact, the displacement field around a fault is explained reasonably well by this type of theoretical model.

Fig. 4.1 illustrates this fault configuration. Fig. 4.1*a* represents the simplest case of a vertical fault plane intersecting the free surface, whereas fig. 4.1*b* shows a more general type – a tilted and buried fault. To describe the configuration, five parameters, $L$, $\theta$, $\delta$, $d$ and $D$, are introduced (see table 4.1).

Measurements of fault movements are always given in relative terms (fig. 4.2). If the relative motion across the fault is predominantly parallel to the fault strike, the fault is called a *strike-slip* (or *transcurrent*) fault. Strike-slip faults are further divided into two slip types, *right-lateral* and *left-lateral*, according to whether the block on one side of the fault moves to the right or left as viewed from the other side. These categories are equally applicable whichever block moves, since the relative displacement of right-(or left-) lateral when seen from one block is also right-(or left-) lateral when seen from the other.

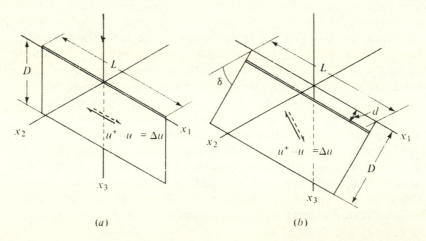

Fig. 4.1. Configuration of fault models for: (*a*) a vertical fault plane intersecting the free surface; and (*b*) a tilted and buried fault plane.

Table 4.1. *Basic parameters of a fault model (static).*

| | |
|---|---|
| Configuration parameters | |
|   Length | *L* |
|   Strike (angle between line of outcrop* and North, for example) | $\theta$ |
|   Dip (angle between direction of steepest slope and horizontal) | $\delta$ |
|   Down-dip depth to the fault's top | *d* |
|   Down-dip depth to the fault's bottom | *D* |
| Offset parameters (see also §4.3.1) | |
|   Displacement vector of a block on one side of the fault relative to the block | |
|   on the other side | $\Delta \mathbf{u}$ or $\mathbf{U}$ |

*Line of surface intersection if the fault is buried.

If the motion is predominantly parallel (or antiparallel) to the fault dip, the fault is called a *dip-slip* fault. Dip-slip faults are classified into two types, *reverse* and *normal*, according to whether the block which lies above the fault (i.e. the hanging-wall block) moves upward or downward relative to the lower block (foot-wall block). A reverse fault may also be called a *thrust* fault, especially if the slip plane makes a low angle with the free surface. If slip on a fault has both dip-slip and strike-slip components of comparable amplitude, it is called an *oblique-slip* fault.

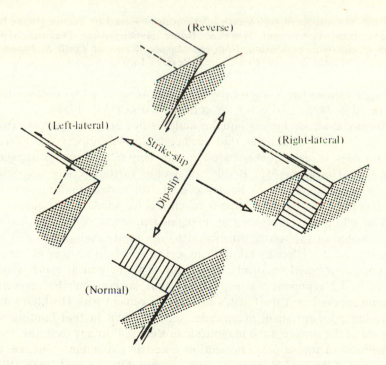

Fig. 4.2. Basic models of fault slipping.

### 4.1.2    Fault statistics

Statistical studies of earthquake faults have been made‘ by several researchers (Tocher, 1958; Iida, 1959, 1965; Press, 1967; Wyss & Brune, 1968) with the aim of establishing empirical formulae for the fault dimension, $L$, as a function of the earthquake magnitude, $M$. These relationships are illustrated in fig. 4.3.

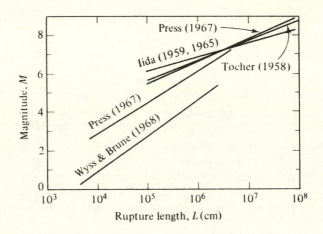

Fig. 4.3. Magnitude–rupture length relations as proposed by Tocher (1958), Iida (1965), Press (1967) and Wyss & Brune (1968). (After Dieterich, 1974. Reproduced, with permission, from the *Annual Review of Earth & Planetary Sciences*, Volume 2, © 1974 by Annual Reviews Inc.)

Fig. 4.4 shows the relation between $L$ and $M$ for fifty-four events which occurred in Japan and other areas in the world (1847–1955).

Surface evidence for earthquake faults is not always apparent. Table 4.2 shows the probabilities that surface faults will appear for different magnitudes of earthquake, based on an empirical study of Japanese events by Hoshino (1956). Briefly, all shallow earthquakes of magnitude 7.4 or more are likely to be associated with surface faulting, but the probability of surface breakage is much less for smaller earthquakes.

This effect can be reasonably explained in terms of the geometrical relationship of the source dimension to the focal depth. Otsuka (1965) has studied this effect by taking a model of spherical sources of various volume, distributed uniformly with depth. The theoretical values shown in table 4.2 represent the probability that spheres for the respective magnitudes intersect the Earth's surface. Agreement with Hoshino's data is rather poor for small magnitude, yet consistent in that faulting will appear at the surface if the magnitude exceeds 7.4. In any case, the linear dimension of the surface intersection takes its maximum value, i.e. the diameter of the sphere, when its depth is zero. Otsuka used Iida's (1959)

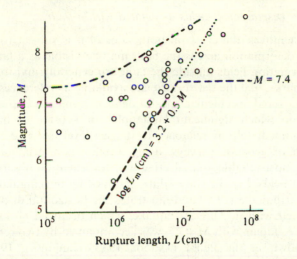

Fig. 4.4. Magnitude–rupture length data (Iida, 1959) interpreted by Otsuka's model (1965). The broken line (log $L_m = 3.2 + 0.5\ M$) gives the upper limit of $L$ (measured in cm) as a function of magnitude, and corresponds to the effective radius of the spherical source. This suggests that the rupture length depends on the focal depth, reaching its maximum value at zero depth. Note that $L$ can be larger than this prediction for magnitudes exceeding 7.4.

data (see fig. 4.4) to introduce a formula for the upper limit of the fault length (cf. (2.23))

$$\log L_m = 3.2 + 0.5M, \tag{4.1}$$

where $L_m$ is in cm. Using the magnitude–energy relation

$$\log E = 11.8 + 1.5M, \tag{4.2}$$

we obtain

$$E \propto L_m^3, \tag{4.3}$$

which supports the source-volume hypothesis that the seismic energy is proportional to the cube of the linear source dimension (§2.3.2).

Table 4.2. *Probability of surface faulting (empirical data from Hoshino, 1956; theoretical data from Otsuka, 1965).*

| Empirical Magnitude | $M < 7.0$ | | | | $7.0 \leq M \leq 7.4$ | | | | $7.4 < M$ | |
|---|---|---|---|---|---|---|---|---|---|---|
| $p(M)$ | 0% | | | | 60% | | | | 100% | |

| Theoretical Magnitude | 5.0 | 5.5 | 6.0 | 6.2 | 6.4 | 6.6 | 6.8 | 7.0 | 7.2 | 7.4 |
|---|---|---|---|---|---|---|---|---|---|---|
| $p(M)$ | 3% | 6% | 11% | 14% | 19% | 25% | 33% | 46% | 66% | 100% |

### 4.1.3   *Displacement fields associated with a fault*

When a fault moves, the crust on both sides of it is subject to deformation. Land deformation across the fault may be visible as a fault offset, but the deformation fields outside the fault are generally invisible, unless stationary marks, like the sea surface, are available as references. Indeed, we have discovered spectacular coastal deformations in occasional seismic events, in which significant land uplift or subsidence have been recognized immediately by reference to the sea level (see §7.2.2).

Repetition of geodetic surveys, using levelling and triangulation, is perhaps the most reliable and effective way for scientific observation of displacement fields. Fig. 4.5 shows data obtained by retriangulation after the Tango earthquake (1927) and illustrates the horizontal displacement field associated with it. This earthquake, which occurred 80 km northwest of Kyoto, Japan, with $M = 7.5$, was accompanied by two remarkable faults as shown in the figure (Tsuboi, 1933; Kanamori, 1973). The principal fault, Gomura, is essentially left-lateral (offset of 2.5 m) with a slight dip-slip component (offset of 0.5 m). The Yamada fault is somewhat minor by comparison.

Fig. 4.5 shows that significant seismic displacements occurred in and around the epicentral area, extending to a distance of 30 km or more from the principal fault. Ignoring the remote fields of more or less irregular features, the vectors in the principal part of the field are seen to lie in a uniform direction, tending to be parallel (or antiparallel) to the fault strike. The northwestward displacements in the Tango Peninsula area increase in amplitude significantly as the fault is approached. The displacement on the other side of the fault is in the opposite direction, i.e. in a left-lateral manner, with offset of 2–3 m. For reference, the inset map in fig. 4.5 (left, bottom) illustrates the radiation pattern (see also fig. 3.1).

Similar studies on many other earthquakes have proved that the orderly arrangement of the displacement vectors and the close association between the earthquake and fault mechanisms are a general characteristic of large earthquakes, although they may be different in detail.

### 4.2   TWO-DIMENSIONAL FAULT MODELS

### 4.2.1   *Elementary considerations*

Reid's hypothesis (1911) explains the earthquake mechanism by a fracture in prestrained crust, which causes 'the elastic rebound of the sides of the fracture towards positions of no elastic strain' (§6.1.1).

The following discussion of fault models refers, in essence, to this

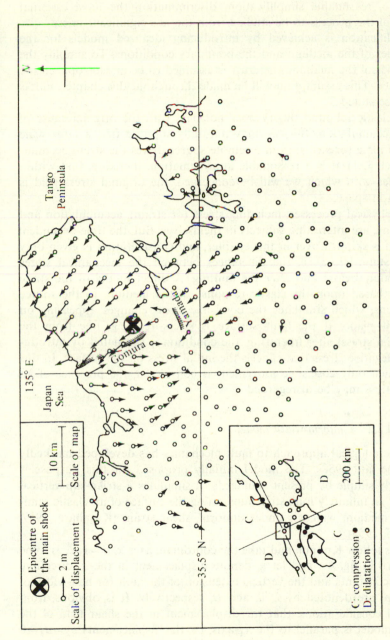

Fig. 4.5. Displacement of triangulation points (Land Survey Department, 1930) in the Tango earthquake of 1927. The distribution of P-wave first motions (Honda, 1932) in the inset shows the radiation pattern. (From Kanamori, 1973. Reproduced, with permission, from the *Annual Review of Earth & Planetary Sciences*, Volume 1. © 1973 by Annual Reviews Inc.)

hypothesis. We may easily imagine that the processes of fracture in the real Earth must be rather complicated. A key to successful modelling is, therefore, reasonable simplification, discriminating the most essential factors of the process to be studied from the less significant ones.

Simplification is achieved by introducing idealized models for the properties of the medium and the boundary conditions. To simplify the properties of the medium, the crust is assumed to be an isotropic elastic half-space. This assumption will be made throughout this chapter, unless otherwise stated.

The elastic rebound theory does not deal with the original cause of strain accumulation. But, we may naturally suppose that a large-scale tectonic force (or stress) from a remote source drives the strain accumulation (cf. §7.1). It is a reasonable approximation, therefore, that within the scales with which we will be concerned, the original stress field is uniform in space.

The physical processes, including stress (or strain), accumulation and fracturing, are interesting subjects in themselves. But, the time-dependent problem is skipped over in this chapter, and consideration is given only to the static effect, i.e. the differences between the mechanical states immediately before and after the faulting. A direct model of the static effect is based upon the elastic response of the medium to the source conditions, which simulates the coseismic stress changes appearing on the fault plane. If the fault surface is stress-free (e.g. free from the tangential stress) after fracturing, the condition is simulated by introducing a negative stress that cancels the initial stress on the surface. In this way, when only coseismic movement is considered, the presence of an initial stress may be disregarded.

### 4.2.2  *Uniform-width model*

The seismological approach to fault mechanics has developed markedly since the late 1950s. The model initially proposed for this purpose is extremely simple by present standards. It simulates a strike-slip vertical fault by an infinitely long strip intersecting the surface of an elastic semi-infinite medium which is under uniform, shear strain (Kasahara, 1957; Knopoff, 1958).

Let us follow Knopoff and take the coordinate axes $x_i$ ($i = 1, 2, 3$) to be as shown in fig. 4.6 and let $u_i$ denote displacement in the $x_i$ direction. Lamé's constants and the vertical extension of the fault (or half-width of the strip) are denoted by $\lambda$, $\mu$, and $D$, respectively. It is obvious, from physical considerations, that the displacement in the shear field of the simple model is parallel to the $x_1$-axis, i.e. the displacements $u_2 = u_3 = 0$ and the operation $(\partial/\partial x_1) = 0$. Therefore, the equations of equilibrium,

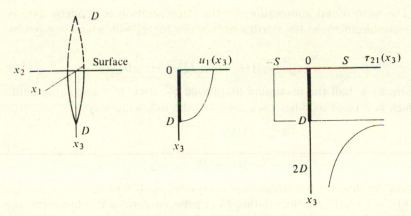

Fig. 4.6. Knopoff's model applied to a vertical strike-slip fault intersecting the surface. Displacement and stress as a function of depth are shown. (After Chinnery, 1967.)

which are in general

$$(\lambda + \mu)(\partial/\partial x_i)\Delta + \mu\nabla^2 u_i = 0 \quad (i = 1, 2, 3), \tag{4.4}$$

take the extremely simple form

$$\mu[(\partial^2/\partial x_2^2) + (\partial^2/\partial x_3^2)]u_1 = 0, \tag{4.5}$$

where

$$\Delta = (\partial u_1/\partial x_1) + (\partial u_2/\partial x_2) + (\partial u_3/\partial x_3),$$

and

$$\nabla^2 = (\partial^2/\partial x_1^2) + (\partial^2/\partial x_2^2) + (\partial^2/\partial x_3^2).$$

The initial field of uniform shear strain, $\tau_{21} = S$, is disturbed by fault formation causing $\tau_{21}$ to become zero at all points on the strip (for $x_2 = 0$ and $|x_3| \leq D$). Knopoff (1958) used an electrical–mechanical analogy to compare the problem with that of the distortion of the electrostatic field by the insertion of a strip conductor. The displacement and stress at $x_2 = 0$ produced by the fault formation are given by

$$u_1(x_3) = (SD/\mu)[1 - (x_3/D)^2]^{1/2} \text{ for } |x_3| \leq D,$$

$$\tau_{21}(x_3) = \begin{cases} -S & \text{for } |x_3| \leq D, \\ S\{[x_3/(x_3^2 - D^2)^{1/2}] - 1\} & \text{for } |x_3| > D, \end{cases} \tag{4.6}$$

as illustrated in fig. 4.6. Equation (4.6) implies that $\tau_{21}$ on the plane $(x_2 = 0, |x_3| \leq D)$, initially $S$, may drop to zero by superposition of this field although significant stress concentration will occur at the edges of the fault.

The most useful information for the interpretation of geodetic data is the displacement on the Earth's surface (i.e. for $x_3 = 0$), which is given by

$$u_1(x_2) = u_m\{[(x_2/D)^2 + 1]^{1/2} - (x_2/D)\}, \qquad (4.7)$$

where $u_m$ is half the maximum amplitude of offset, $U_m$, across the fault, which is related to other parameters in the following way:

$$U_m = 2SD/\mu,$$
$$\Delta\sigma = S = \tfrac{1}{2}(U_m\mu/D), \qquad (4.8)$$

where $\Delta\sigma$ denotes the stress drop associated with faulting.

Fig. 4.7 illustrates the relation (4.7) between surface displacement, $u_1$, and the dimensionless distance $x_2/D$. Fig. 4.8a shows the displacement field about a strike-slip fault.

Starr (1928) proposed a two-dimensional elliptical crack model in an infinite medium. However, it took a considerable time for seismologists to notice the usefulness of his model for the study of dip-slip faults. Basically, his model is of an elliptical crack subject to a shear stress $\tau_{23} = S$, applied at $x_2 = \infty$, causing a displacement of the sides of the crack in the $x_3$-direction. The solution for a dip-slip fault of infinite length is obtained from Starr's model by setting the minor axis of his ellipse equal to zero (fig. 4.8b). Thus the displacement at $x_2 = 0$ is given by

$$u_3(x_3) = \tfrac{1}{2}SD[(\lambda + 2\mu)/\mu(\lambda + \mu)][1 - (x_3/D)^2]^{1/2} \quad \text{for } 0 \leq x_3 \leq D, \qquad (4.9)$$

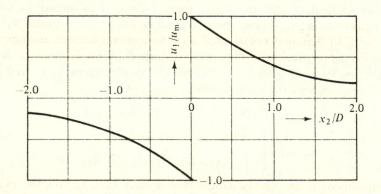

Fig. 4.7. Surface displacement parallel to fault ($u_1$) as a function of distance from the fault ($x_2$) for the vertical strike-slip fault model shown in fig. 4.8a. In this figure, $u_1$ and $x_2$ are given in normalized units, taking $u_m$ and $D$ respectively as normalizations.

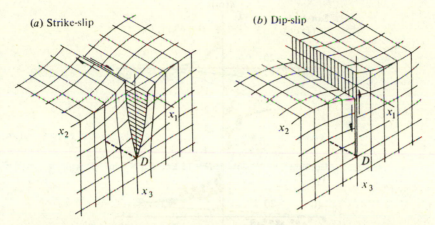

Fig. 4.8. Displacement fields associated with vertical fault models for: (*a*) strike-slip; and (*b*) dip-slip. Displacements are exaggerated. Also, the vertical scale is exaggerated by a factor of about four, so the crack is, in fact, much narrower than it appears in the drawing.

where $D$ denotes the half-width of the crack. The maximum offset occurs at the centre of the crack, and, if $\lambda = \mu$, is given by

$$U_m = \tfrac{3}{2}(SD/\mu),$$

or, (4.10)

$$\Delta\sigma = S = \tfrac{2}{3}(U_m\mu/D).$$

These two models of strike-slip and dip-slip faults are similar to one another, in their functional forms, as can be seen from a comparison of (4.6) and (4.9) as well as (4.8) and (4.10). For example, the functional form of $u_3(x_3)$ in Starr's model (4.9) is similar to that for $u_1(x_3)$ in Knopoff's model (4.6). It must be remarked, however, that $\tau_{31}$ is not equal to zero on $x_3 = 0$ in Starr's model, and therefore, strictly speaking, this model does not apply immediately to surface faulting.

### 4.2.3   Determination of fault parameters

Application of the two-dimensional models to actual faults is not always justified. It is evident in fig. 4.5 that the displacement field about the edges of the fault is greatly complicated by the finiteness of the fault. Nevertheless, the two-dimensional model is often used, for simplicity's sake, to interpret the field data. The following are examples of this kind of analysis, in which only data from the stations located in the central part of the displacement field are used so that the two-dimensional aspects of the fault may be emphasized.

Fig. 4.9 illustrates the displacement ($u$) versus distance ($x_2$) relation in

64 *Earthquake faults*

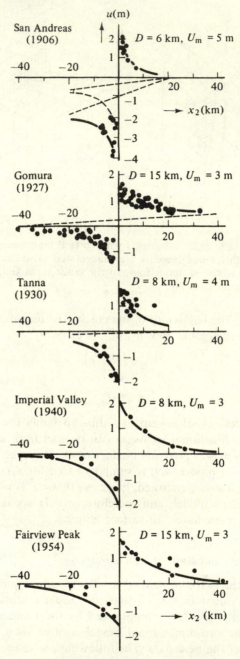

Fig. 4.9. Surface displacements (parallel to the fault strike) associated with several representative strike-slip faults for comparison with the predictions of the vertical strike-slip model (fig. 4.7). Broken lines show corrections for the hypothetical strain accumulation during the period between pre- and post-earthquake surveys. (After Kasahara, 1960.)

five earthquake faults of strike-slip type. Data from stations situated on fault edges are not used for the reason stated above. In order to compare the theoretical curve (fig. 4.7) with the observations, the dimensionless coordinates, $x_2/D$ and $u_1/u_m$, in fig. 4.7 must be converted to $x_2$ and $u_1$, respectively, by estimating $D$ and $u_m$. The two parameters $D$ and $u_m$ are adjusted in a trial and error manner until the best fit of the theoretical curve with the observations is found. In the case of the Tango earthquake, for example, we obtain

$$D = 15 \text{ km}, \quad U_m = 2u_m = 3 \text{ m}, \tag{4.11}$$

which yields the theoretical curve shown in fig. 4.9 (second from the top). This technique for the determination of parameters $D$ and $u_m$ has been applied to four other examples, which are also shown in fig. 4.9. Broken lines in the upper three diagrams represent hypothetical strain accumulation during the period between the preseismic and postseismic surveys. The data should be adjusted to account for this before the technique described above is used to determine the fault parameters (Kasahara, 1958b).

Substituting the parameters into (4.8), we are able to estimate $S$ (more precisely, the stress drop $\Delta\sigma$) on the basis of the present model. For example, the stress drop in the Gomura fault is estimated as,

$$\Delta\sigma = 50 \text{ bar}, \tag{4.12}$$

if $\mu = 5 \times 10^{11}$ dyn cm$^{-2}$. Further consideration of these parameters enables us to estimate the strain energy change, $E_f$, if the fault length can be estimated. The main parameters associated with earthquakes are listed in appendix 1 for a large number of data sources.

We have developed the discussion of fault mechanics on the basis of simple models. As will be seen later, however, mathematical techniques are now available for studying more advanced models. By present standards, therefore, the classical models given above are extremely simple. Yet, they are useful for quick interpretation of data, especially when the quality or the quantity of data does not permit precise analysis.

## 4.3   THE ELASTICITY THEORY OF DISLOCATION

The scope of fault studies has been widened remarkably by the introduction of dislocation theory into this research field. As stated previously (§4.2.2), the concept of dislocation itself is not new in metallurgy and crystal physics. However, the idea of dislocation has only received attention in the field of earthquake sciences since the late 1950s.

Steketee (1958a, b), referring to the classic work of Volterra (1907), constructed the mathematical foundations for the study of a three-dimensional crack, or fault. Because his model was based on the same foundation as crystal dislocation theory, and because it dealt with continuous dislocation in contrast to the discrete dislocation of the classical theory, Steketee carefully distinguished it by calling it the *elasticity theory of dislocations*. His work is especially notable because it founded the basis for the application of dislocation theory to seismological problems.

### 4.3.1    *A surface of shear dislocation*

A dislocation surface is, briefly, a surface of discontinuity in displacement. It may be understood conceptually by visualizing its formation in the following process (fig. 4.10):

(*a*) make a cut, $\Sigma$, inside an elastic body, to form an arbitrary new surface;

(*b*) apply relative displacement of slip-type to the two faces of the cut;

(*c*) rejoin the faces in their new positions.

In this way the body regains its stress continuity but has spontaneous internal strains due to the discontinuous displacement across $\Sigma$. This type of deformation is called *shear dislocation*. The strain pattern in the

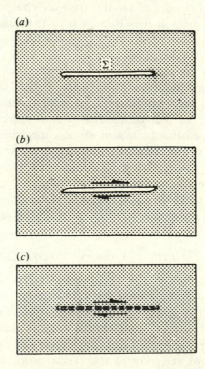

Fig. 4.10. Formation of a shear dislocation surface: (*a*) cutting; (*b*) applying relative displacement, or slip; and (*c*) rejoining.

medium depends on two factors, the configuration of the surface and the offset distribution on it.

Let $\Sigma^+$ and $\Sigma^-$ denote the two new faces after cutting $\Sigma$. Take an arbitrary point $P$ in $\Sigma$, and denote its displacement on $\Sigma^+$ and $\Sigma^-$ as $u^+$ and $u^-$, respectively, Then,

$$\Delta u = u^+ - u^- \qquad (4.13)$$

represents the relative displacement of the two faces at $P$, or the discontinuous displacement across $\Sigma$ at $P$. Sometimes, the symbol $[u]$ is used instead of $\Delta u$. It may also be written as $\Delta \mathbf{u}$, or $(\Delta u_1, \Delta u_2, \Delta u_3)$, or $[u_k]$, where $k = 1, 2, 3$, so that it can be recognized as a vector.

For mathematical purposes, an appropriate displacement discontinuity may be assumed on $\Sigma$. A natural distribution may be like the pattern in fig. 4.11, with dislocation amplitude decreasing gradually to zero at the surface edges. If we assumed uniform dislocation of $\Delta u$, as is often the case for simple fault models, we would have an unnatural overlapping of the medium at the edges. This sort of physical paradox may be avoided, in practice, by hypothesizing a small hole at each edge.

The original crystallographic models of dislocation are discrete in character (e.g. Thomson & Seitz, 1968). In a structure such as a crystal lattice, a dislocation occurs only in a restricted manner, causing lattice offset discretely in units of the lattice constant. This mode of lattice slip, of uniform amplitude, is represented by a vector $\mathbf{b}$, which is called the *Burgers vector*. The dislocation is called either *edge type* or *screw type*, according to whether the direction of the Burgers vector is parallel to or perpendicular to the direction of propagation of the dislocation area respectively (see Bilby & Eshelby, 1968, for further discussion). Therefore, the strike-slip (*a*) and the dip-slip (*b*) fault types shown in fig. 4.8 correspond to the screw and the edge types, respectively.

The concept of dislocation has been generalized so that it applies to seismological problems. As will be seen later, an earthquake fault, when considered as a huge dislocation in the Earth, has marked differences from a dislocation in a laboratory specimen, in terms of the dimensions and the amplitude of offset. More essentially, the seismic dislocation cannot be discrete as in crystalline materials. In other words, the subject

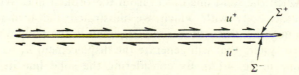

Fig. 4.11. A shear dislocation surface in an elastic body, and the displacements along it.

of the following discussion is continuous dislocation, although terms and formulae may appear to be similar to those used in crystallography.

### 4.3.2   *The force–dislocation equivalence*

A dislocation surface of the type described above causes strains around it. In this case, there is no external force exerted on the medium. The question of whether it is possible to find a system of external forces which generates the same strains in the medium without the introduction of a cut then arises. The answer is affirmative within the infinitesimal strain theory of elasticity. The theorem of *force–dislocation equivalence* states that the displacement field generated by a shear dislocation on an infinitesimal surface element is the same as that generated by a double-couple applied at the surface element in the absence of dislocation. The proof of the theorem can be found, for example, in Steketee (1958*b*). A mathematically neat presentation of the equivalence is given in Landau & Lifshitz (1970). In the present section, however, we shall follow a conceptual explanation by Maruyama (1973).

Let us consider shear dislocation at a small plane element $\Sigma$, in an isotropic elastic body, where $\Sigma$ is taken to be perpendicular to the $x_2$-axis, and where the origin of the rectangular Cartesian coordinate system is included in $\Sigma$. The relative displacement of the medium across $\Sigma$ is illustrated schematically in fig. 4.12*a* (broken lines) with $u_1^+$ and $u_1^-$ on the $\Sigma^+$ and $\Sigma^-$ sides, respectively. We assume that only the displacement component $u_1$ is discontinuous and that both $u_2$ and $u_3$ are continuous across $\Sigma$.

One may doubt that the abrupt change of displacement, as shown by the broken line in the figure, may be generated by a force system without the introduction of a cut. Intuitively, a discontinuity of displacement cannot be replaced by any force system if the force system does not break the material. But we should remember that we are considering the problem from within the framework of the idealized linear theory of elasticity.

Now, the broken line in fig. 4.12*a* can be taken as the limit, as $\varepsilon$ approaches zero, of a smooth variation of displacement $u_1$ (shown by the solid line in fig. 4.12*a*), where $\varepsilon$ denotes a short distance from the element $\Sigma$ within which the solid line differs from the broken line. We will now reconsider the body with which we illustrated the formation of a dislocation in §4.3.1. Without introducing a cut in the body, we will seek a body force system that will generate the displacement $u_1$ as shown by the solid line in fig. 4.12*a*. By considering the solid line in the figure instead of the broken line, the variation of $u_1$ at the edge of $\Sigma$ can be adequately modelled. In the vicinity of every edge of $\Sigma$, the absolute values of displacement $u_1^+$ *and* $u_1^-$ as shown in fig. 4.12*a* are assumed to

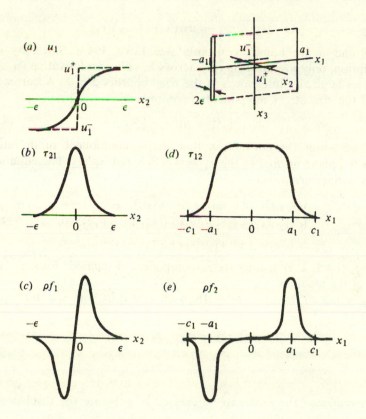

Fig. 4.12. Schematic view of the surface of a shear dislocation (right, top) and elastic deformation fields (a)–(e). (a) Abrupt (broken line) and smooth (solid line) displacement change across the surfaces of a shear dislocation. (b) and (d) represent the stresses, and (c) and (e) represent the forces that will cause the smooth variation shown in (a). The diagram at the top right shows the coordinate system and parameters associated with the dislocation. (After Maruyama, 1973.)

be smoothly and uniformly diminished as we proceed outward from $\Sigma$. The stress–strain relations in an isotropic elastic body are given by

$$\tau_{11} = \lambda \Theta + 2\mu(\partial u_1/\partial x_1),$$
$$\tau_{22} = \lambda \Theta + 2\mu(\partial u_2/\partial x_2),$$
$$\tau_{33} = \lambda \Theta + 2\mu(\partial u_3/\partial x_3),$$

$$\tag{4.14}$$

$$\tau_{23} = \tau_{32} = \mu[(\partial u_3/\partial x_2) + (\partial u_2/\partial x_3)],$$
$$\tau_{31} = \tau_{13} = \mu[(\partial u_1/\partial x_3) + (\partial u_3/\partial x_1)],$$
$$\tau_{12} = \tau_{21} = \mu[(\partial u_2/\partial x_1) + (\partial u_1/\partial x_2)],$$

where

$$\Theta = (\partial u_1/\partial x_1) + (\partial u_2/\partial x_2) + (\partial u_3/\partial x_3),$$

and $\lambda$ and $\mu$ are Lamé's constants (see Love, 1944). Since, by our assumption, only $u_1$ varies sharply across $\Sigma$, only the derivative $\partial u_1/\partial x_2$ possesses large absolute values in the neighbourhood of $\Sigma$. Accordingly, two of the nine stress components predominate:

$$\tau_{12} = \tau_{21} = \mu(\partial u_1/\partial x_2). \tag{4.15}$$

Here, we adopt the convention that the $x_2$-component of the stress across the plane normal to the $x_1$-axis is denoted by $\tau_{12}$. The equations of equilibrium are

$$(\partial \tau_{11}/\partial x_1) + (\partial \tau_{21}/\partial x_2) + (\partial \tau_{31}/\partial x_3) + \rho f_1 = 0,$$
$$(\partial \tau_{12}/\partial x_1) + (\partial \tau_{22}/\partial x_2) + (\partial \tau_{32}/\partial x_3) + \rho f_2 = 0, \tag{4.16}$$
$$(\partial \tau_{13}/\partial x_1) + (\partial \tau_{23}/\partial x_2) + (\partial \tau_{33}/\partial x_3) + \rho f_3 = 0,$$

where $f_k$ ($k = 1, 2, 3$) denotes the $x_k$-component of the body force per unit mass of the medium.

Neglecting all the terms other than those that include $\tau_{12}$ and $\tau_{21}$, we find

$$\rho f_1 = -(\partial \tau_{21}/\partial x_2),$$
$$\rho f_2 = -(\partial \tau_{12}/\partial x_1), \tag{4.17}$$
$$\rho f_3 = 0,$$

which represent the forces to be exerted to generate the displacement field $u_1$.

Fig. 4.12*b* shows the variation of $\tau_{21}$ along the $x_2$-axis, which is derived from (4.15) using the function $u_1(x_2)$ shown by the solid line in fig. 4.12*a*. The variation of $\tau_{12}$ along the $x_1$-axis is shown in fig. 4.12*d*, where the plane element $\Sigma$ is assumed to have its edge at $x_1 = \pm a_1$. The plateau in this curve has the same height as the curve in fig. 4.12*b*, since the variation of $u_1$ along a line which is parallel to the $x_2$-axis is assumed to be the same as the curve in fig. 4.12*a*, as long as the line does not extend into the vicinity of the edge. The decrease in the ordinate near each edge in fig. 4.12*d* is naturally supposed to be smooth, and we have $\tau_{12} = 0$ for $|x_1| \geqq c_1$, $c_1$ being assumed to differ from $a_1$ only by a very small quantity. The curves in figs. 4.12*c* and *e* are derived from the $\tau_{21}$ and $\tau_{12}$ curves (shown in figs. 4.12*b* and *d*) using (4.17).

As can be seen in figs. 4.12*c* and *e*, the forces $\rho f_1$ and $\rho f_2$ have moment about the $x_3$-axis. Consider first the forces to be exerted in the $x_1$-direction. Since $\rho f_1$ represents the body force per unit volume, the moment due to the forces in the $x_1$-direction per unit area of the surface

$\Sigma$ is given by

$$-\int_{-\varepsilon}^{\varepsilon} x_2 \rho f_1 dx_2 = \int_{-\varepsilon}^{\varepsilon} x_2 (\partial \tau_{21}/\partial x_2) dx_2$$

$$= [x_2 \tau_{21}]_{-\varepsilon}^{\varepsilon} - \int_{-\varepsilon}^{\varepsilon} \tau_{21} dx_2$$

$$= -\mu \int_{-\varepsilon}^{\varepsilon} (\partial u_1/\partial x_2) dx_2 = -\mu \Delta u_1. \qquad (4.18)$$

Similarly, the moment per unit length in the $x_3$-direction due to the forces to be exerted in the $x_2$-direction is written as

$$\int_{-\varepsilon}^{\varepsilon} dx_2 \int_{-c_1}^{c_1} x_1 \rho f_2 dx_1 = -\int_{-\varepsilon}^{\varepsilon} dx_2 \int_{-c_1}^{c_1} x_1 (\partial \tau_{12}/\partial x_1) dx_1$$

$$= \int_{-\varepsilon}^{\varepsilon} dx_2 \left\{ -[x_1 \tau_{12}]_{-c_1}^{c_1} + \int_{-c_1}^{c_1} \tau_{12} dx_1 \right\}$$

$$= \int_{-\varepsilon}^{\varepsilon} dx_2 \int_{-c_1}^{c_1} \mu(\partial u_1/\partial x_2) dx_1$$

$$= \mu \int_{-c_1}^{c_1} dx_1 \int_{-\varepsilon}^{\varepsilon} (\partial u_1/\partial x_2) dx_2$$

$$= \mu \int_{-c_1}^{c_1} \Delta u_1 dx_1. \qquad (4.19)$$

This shows that the forces to be exerted in the $x_2$-direction have an average moment $\mu \Delta u_1$ per unit area of $\Sigma$. It follows that the forces to be exerted in the $x_1$- and $x_2$-directions are associated with moments of equal magnitude and of opposite sign about the $x_3$-axis. A formula is thus established for the component moment of the forces to be exerted,

$$M_0 = \mu U A, \qquad (4.20)$$

where $\mu$ is the rigidity of the medium and $U$ is the average of $\Delta u_1$ over the surface $\Sigma$ which has a total area $A$.

Now let us consider the case in which we diminish both $\varepsilon$ and the linear dimensions of $\Sigma$ indefinitely, keeping the origin of the coordinate system in $\Sigma$ and keeping $\Delta u_1$ at the assigned value. Then, the form of the displacement $u_1$, shown by the solid line in fig. 4.12$a$, approaches the broken line and the curves in figs. 4.12$b$ and $d$ become taller and thinner. As we approach the limit, the force system shown in figs. 4.12$c$ and $e$ tends to a double-couple, each component couple having moment $\mu \Delta u_1$ per unit area of $\Sigma$. Note that in this derivation we have employed the assumption that the stress–strain relations with constant coefficients (4.14) hold everywhere, even at the limit.

We thus arrive at the statement that *the displacement field due to a shear dislocation is mathematically expressed by a surface integral (super-position) of point-source elements of the double-couple type over the entire fault surface.*

### 4.3.3    *Three-dimensional fault models*

Let a point $P(\xi_1, \xi_2, \xi_3)$ in the medium be subject to a single force of magnitude $F$ in the direction $x_k$ $(k = 1, 2, 3)$. The displacement field thus generated is represented generally by $u_{mk}(Q, P)$, which denotes the $x_m$-component $(m = 1, 2, 3)$ of the displacement at a point $Q(x_1, x_2, x_3)$.

If $P$ is subject to two equal and opposite forces of magnitude $F/2h$ separated by a distance $2h$ as shown in fig. 4.13, the displacement at $Q$ ($x_m$-component) is given by,

$$u_m(Q) = (1/2h)[u_{m2}(Q, \xi_1 + h, \xi_2, \xi_3) - u_{m2}(Q, \xi_1 - h, \xi_2, \xi_3)]. \quad (4.21)$$

This approaches $\partial u_{m2}(Q, P)/\partial\xi_1$ as $h \to 0$ and the force system at $P$ has a moment equal to $F$.

If the force system is of the double-couple type with a component couple of magnitude $M_0$ (see figs. 4.12c and e), then the displacement at $Q$ is given by

$$u_m(Q) = \frac{\partial}{\partial\xi_1} u_{m2}(Q, P) + \frac{\partial}{\partial\xi_2} u_{m1}(Q, P), \quad (4.22)$$

where $M_0$ replaces the moment $F$. Equation (4.22) is used to derive the displacement field due to a dislocation surface $\Sigma$, by replacing $M_0$ by $\mu\Delta u_1 d\Sigma$ (cf. (4.20)) and integrating (4.22) over the surface $\Sigma$ (see Maruyama, 1973, for a detailed discussion).

The displacement component $u_{mk}(Q, P)$ in (4.21) and (4.22) is generally

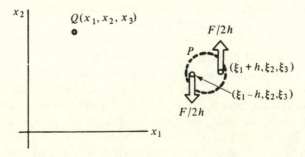

Fig. 4.13. A force couple at a point $P$, causing a displacement at the point $Q$.

expressed as (Maruyama, 1973):

$$u_{mk}(Q,\ P)=\frac{F}{4\pi\mu}\left[\delta_{mk}\frac{1}{r}-\frac{\lambda+\mu}{2(\lambda+2\mu)}\frac{\partial^2 r}{\partial x_m \partial x_k}\right],\qquad(4.23)$$

where $\delta_{mk}$ is the delta function ($\delta_{mk}=1$ if $m=k$, and $\delta_{mk}=0$ if $m\neq k$), and $r$ denotes the distance between $P$ and $Q$, i.e.

$$r=(r_1^2+r_2^2+r_3^2)^{1/2},\quad r_k=x_k-\xi_k\quad(k=1,\ 2,\ 3).\qquad(4.24)$$

Substituting (4.23) into (4.22), we can obtain an expression for the displacement field due to shear dislocation.

More generally, (4.23) enables us to write the displacement at $Q$ due to a dislocation surface with arbitrary displacement vector $\Delta u_k (k=1, 2, 3)$ as

$$u_m(Q)=\iint_\Sigma \Delta u_k(P)T_{kl}^m(P,\ Q)v_l \mathrm{d}\Sigma,\qquad(4.25)$$

where $v_l (l=1,\ 2,\ 3)$ is the normal to the surface $\Sigma$ at $P(\xi_1,\ \xi_2,\ \xi_3)$ (where the direction from $\Sigma$ to $\Sigma^+$ is regarded as positive), and,

$$\Delta u_k(P)=u_k^+ - u_k^- \quad(k=1,\ 2,\ 3),\qquad(4.26)$$

$$T_{kl}^m=\frac{1}{4\pi}\left[\frac{\mu}{\lambda+2\mu}\left(-\delta_{kl}\frac{r_m}{r^3}+\delta_{mk}\frac{r_l}{r^3}+\delta_{ml}\frac{r_k}{r^3}\right)+3\left(\frac{\lambda+\mu}{\lambda+2\mu}\right)\frac{r_k r_l r_m}{r^5}\right]\qquad(4.27)$$

and the surface element of $\Sigma$ is denoted by $\mathrm{d}\Sigma$. Equation (4.25) is sometimes called *Volterra's theorem* (Volterra, 1907) as it was first derived by him to describe the displacement field due to a dislocation of a particular type (Volterra dislocation). A derivation of (4.25) may be found in Steketee (1958a), where Betti's reciprocal theorem is applied to a body which has an internal dislocation surface $\Sigma$. Alternatively, the displacement field due to a distribution of force nuclei, which is equivalent to the discontinuous movement across $\Sigma$, may be calculated (see Landau & Lifshitz, 1970).

The displacement field due to a dislocation in a semi-infinite medium may be derived from (4.25) to (4.27) which are applicable to an infinite medium. Let the surface, $S$, of the semi-infinite medium be the plane $x_3=0$, where the positive $x_3$-axis penetrates the medium (fig. 4.14). Then, the surface must be free from the stresses $\tau_{13}$, $\tau_{23}$ and $\tau_{33}$, i.e.

$$\tau_{13}=\tau_{23}=\tau_{33}=0 \quad\text{at } x_3=0.\qquad(4.28)$$

To obtain a solution, we first take an infinite medium and introduce a set of forces in it as follows:
(a) a double force (kl) at $P$;
(b) a double force (kl)' at $P'$;
(c) a load normal to the plane $x_3=0$ (see below);

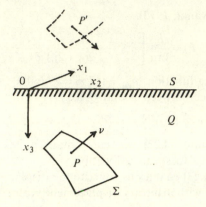

Fig. 4.14. A dislocation surface in a semi-infinite medium. $P'$ is the image point of $P$ (see text).

with $k$, $l$ fixed. The force (*a*) above represents the force (i.e. *nucleus of strain*) developed by the supposed dislocation at $P(\xi_1, \xi_2, \xi_3)$, whereas the force (*b*) is introduced at $P'(\xi_1, \xi_2, -\xi_3)$ which is the image point of $P$. The force (*b*) is chosen so that the tangential stresses $\tau_{13}$ and $\tau_{23}$ from the forces (*a*) and (*b*) may be cancelled on the plane $x_3 = 0$. As a result, the stress normal to the plane $x_3 = 0$ is double that from (*a*) alone. Therefore we consider the force (*c*) as equal and opposite to the stress normal to the plane $x_3 = 0$ due to forces (*a*) and (*b*). It is clear that the resultant field satisfies the boundary conditions given in (4.28) and that it vanishes at infinity if the proper form for force (*c*) is selected. Let us denote the $m$-component of the resultant displacement due to the forces (*a*), (*b*) and (*c*) for fixed $k$ and $l$ by $W_{kl}^m$, then we obtain the displacement field in the semi-infinite medium as

$$u_m(Q) = \iint \Delta u_k(P) W_{kl}^m (P, Q) \nu_l d\Sigma, \qquad (4.29)$$

where $T_{kl}^m$ in (4.25) is replaced by $W_{kl}^m$. $W_{kl}^m$ represents the displacement field of a set of nuclei of strain of the $k,l$-type acting at an internal point $P$ in a semi-infinite elastic medium.

### 4.3.4   *Displacement fields on the free surface*

Evaluation of (4.29) would, in general, require elaborate mathematics. In other words, it is difficult to apply to practical problems unless some convenient formulae are derived from it for efficient numerical calculation. Several notable papers have been published dealing with this problem (e.g. Chinnery, 1961; Maruyama, 1964; Press, 1965; Savage &

Hastie, 1966; Mansinha & Smylie, 1971; Matsu'ura & Sato, 1975), and it is now possible to apply advanced fault models to a wide variety of cases.

The information which concerns us most is undoubtedly the surface displacement field about a fault. Let us study this problem, in the first instance, following Chinnery (1961). He assumed a vertical strike-slip fault buried at a depth $d$ from the free surface ($d = 0$ in fig. 4.1$a$), and then derived from (4.29) the following set of expressions for the displacement at a surface point ($x_1$, $x_2$, 0),

$$\frac{u_1}{U} = -\frac{1}{8\pi}\left( x_2 z \frac{3s+4y_3}{s(s+y_3)^2} - 4\tan^{-1}\frac{x_2 s}{y_3 z}\right)\Big\|,$$

$$\frac{u_2}{U} = \frac{1}{8\pi}\left( \ln(s+y_3) + \frac{y_3}{s+y_3} - x_2^2\frac{3s+4y_3}{s(s+y_3)^2}\right)\Big\|, \qquad (4.30)$$

$$\frac{u_3}{U} = \frac{x_2}{4\pi}\left( \frac{s+2y_3}{s(s+y_3)}\right)\Big\|.$$

It is assumed that ($y_1$, 0, $y_3$) represent the coordinates of a point on the fault surface, and $\lambda = \mu$ and that fault slip ($U = \Delta u$) is uniform (the coordinate system and parameters refer to fig. 4.1). In (4.30), the double vertical line represents the following operation,

$$f(y_1, y_3)\| = f(L/2, D) - f(L/2, d) - f(-L/2, D) + f(-L/2, d), \quad (4.31)$$

and,

$$z = x_1 - y_1,$$
$$s^2 = z^2 + x_2^2 + y_3^2. \qquad (4.32)$$

Therefore, the displacement ($u_1$, $u_2$, $u_3$) at a surface point may be calculated by first evaluating the functions in (4.30) at the respective corner point of the rectangular fault surface, and then carrying out the operation (4.31) (fig. 4.15$a$ shows an example of displacements due to a fault of this type).

Further investigation of a vertical fault model has been carried out by Press (1965), who has calculated the surface distribution of various displacement and strain (tilt) components due to strike-slip and dip-slip fault surfaces of typical aspect ratios ($L/D = 2/1$ and 20/1, with $d = 0$). A set of formulae by Mansinha & Smylie (1971) are often used for a tilted fault such as that shown in fig. 4.1$b$. Matsu'ura & Sato (1975) further refined the theory for strike-slip and dip-slip models, and fig. 4.15 shows their results for the surface displacement fields of two typical fault models.

Fig. 4.15$a$ applies to a vertical strike-slip fault, as discussed by Chinnery (see (4.30)–(4.32)), and illustrates vertical (left) and horizontal (right) surface displacements. The unit of displacement is $U/100$, the aspect

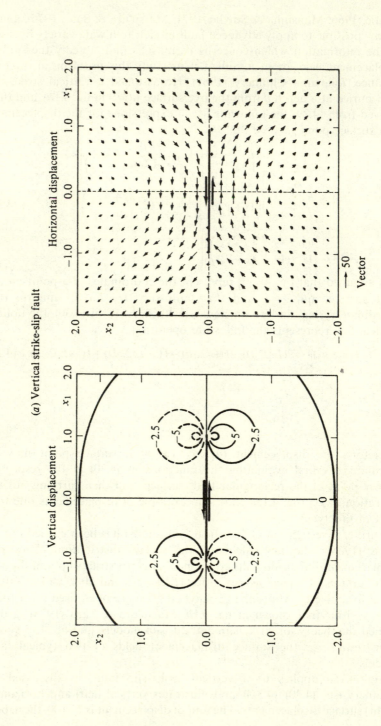

(a) Vertical strike-slip fault

(b) Tilted dip-slip fault

Vertical displacement

Horizontal displacement

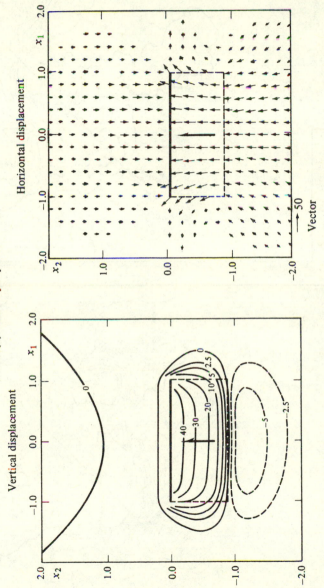

Fig. 4.15. Predicted surface displacements for: (a) a vertical strike-slip model; and (b) a tilted dip-slip model. The left-hand diagrams show vertical and the right-hand diagrams horizontal displacements. The unit of displacements is $U/100$ (i.e. percentage of the fault slip) and the aspect ratio is 2/1. (From Matsu'ura & Sato, 1975.)

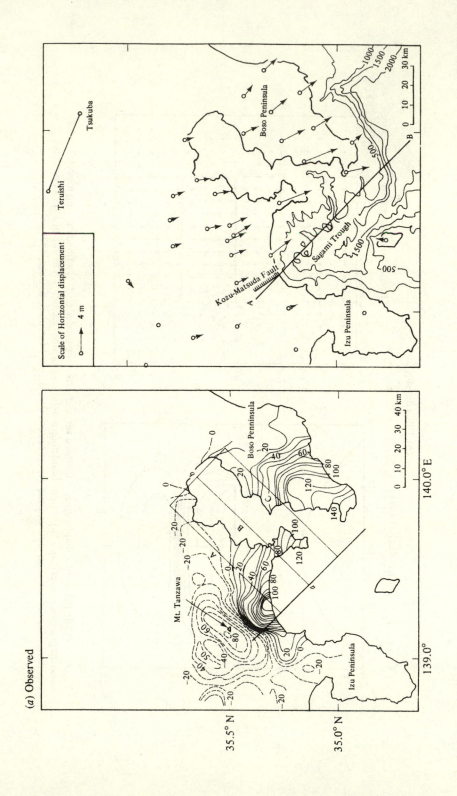

(a) Observed

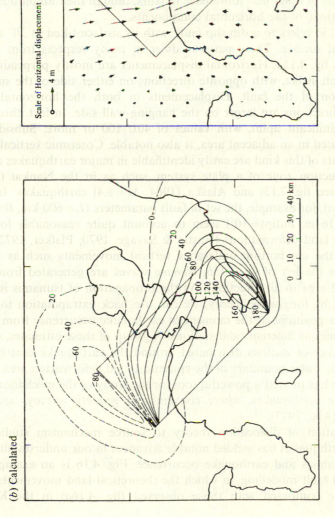

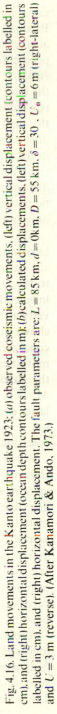

Scale of Horizontal displacement

4 m

(b) Calculated

Fig. 4.16. Land movements in the Kanto earthquake 1923: (a) observed coseismic movements. (left) vertical displacement (contours labelled in cm), and (right) horizontal displacement (ocean depth contours labelled in m); (b) calculated displacements. (left) vertical displacement (contours labelled in cm), and (right) horizontal displacement. The fault parameters are: $L = 85$ km, $d = 0$ km, $D = 55$ km, $\delta = 30$°, $U_s = 6$ m (right-lateral) and $U = 3$ m (reverse). (After Kanamori & Ando, 1973.)

ratio (given by $L/(D-d)$ for a buried fault) is 2/1 and $d$ is $L/30$. Predominance of horizontal displacement reflects strike-slip movement at the fault. In a zone crossing the central part of the fault perpendicularly, vectors of displacement appear almost parallel to the fault strike. If the amplitudes of these vectors are plotted against distance from the fault $(x_2)$, then a distribution curve like that shown in fig. 4.7 for the two-dimensional case is obtained. If, on the other hand, a wider view of the field is taken, the vectors approximate a radial pattern at larger distances from the fault. The geodetic aspect of the Gomura fault shown in fig. 4.5 can be well explained by this pattern, if it is compared to the right half of the vertical strike-slip model shown in fig. 4.15a. Vertical displacements are greatest at the extremities of the fault, with characteristic patterns of uplift and subsidence in pairs, though their amplitudes are only fractions of the horizontal components.

Fig. 4.15b refers to a dip-slip fault, with its surface tilted at 30° to the horizontal surface. These patterns differ in many respects from those shown in fig. 4.15a. Horizontal displacements are mostly perpendicular to the fault strike, with opposite directions on either side of the surface intersection of the fault. Displacements in both the horizontal and vertical directions are larger on the hanging-wall side. In fact, this area shows significant uplift, with values of $40U/100$ or more. Subsidence, concentrated in an adjacent area, is also notable. Coseismic vertical land movements of this kind are easily identifiable in major earthquakes along the subduction zone of a plate system, such as in the Nankai (1946, $M = 8.1$; see fig. 7.13) and Alaska (1964, $M = 8.4$) earthquakes. In the latter event, for example, the set of fault parameters ($L = 600$ km, $D = 200$ km, $U = 16$ m, $\delta(\text{dip}) = 10°$) seem to account quite reasonably for the coseismic land deformations (Hastie & Savage, 1970; Plafker, 1972).

When the sea bottom undergoes vertical movements such as those shown on the left of fig. 4.15b, *tsunami* waves are generated from the epicentral area in all directions. As the propagation of tsunamis is well described by long-wave theory, we can use back extrapolation to estimate their positions at the moment of earthquake occurrence from their arrival times at different tidal stations. A group of these estimates, made at a number of stations distributed around the earthquake centre, will suggest the outer boundary of the epicentral land deformation area. This technique has proved a powerful tool for the study of the mechanisms of submarine earthquakes, where conventional geodetic surveys are impossible (Abe, 1973).

Application of dislocation theory to source mechanism studies of major earthquakes has yielded notable advances in our understanding of fault dynamics and earthquake occurrence. Fig. 4.16 is an example of successful fault modelling, in which the theoretical land movements (fig. 4.16b) are compared with those observed (fig. 4.16a) in the Kanto earthquake, Japan (1923, $M = 7.9$) (Ando, 1971; Kanamori & Ando, 1973). This is an oblique-slip tilted fault along a submarine trough in the Sagami Bay, and the theoretical patterns are drawn by linear combination of the displacement fields produced by the strike-slip and dip-slip components of the fault. In a trial and error manner, Kanamori & Ando found the following set of parameters to give the best fit to the observed patterns:

$$L = 85 \text{ km}, \quad d = 0 \text{ km}, \quad D = 55 \text{ km}, \quad \delta = 30°, \quad \text{strike N45°W},$$
$$U_0(\text{right-lateral}) = 6 \text{ m}, \quad U(\text{reverse}) = 3 \text{ m}.$$

Notice, in fig. 4.16*b*, that the theoretical model explains the observed coseismic land movements beautifully, in terms of both vertical and horizontal components. Let us call this model the geodetic model. Source mechanism studies of this earthquake, which are based upon seismic wave radiation and are independent of the geodetic aspect, also indicate a tilted fault ($\delta = 34°$), striking N70°W and undergoing oblique (right-lateral and reverse) slip for 2 m (Kanamori & Ando, 1973). Good agreement between the geodetic and seismic models, as achieved in the example shown in fig. 4.16, proves the validity of this technique for source mechanism study (quantitative disagreement of the slip amplitude in the two models is a problem for future study). Several more examples where good agreement has been achieved between the two kinds of earthquake model are given in appendix 1.

# 5 Moving dislocations

TELESEISMIC EVIDENCE FOR RUPTURE PROPAGATION

The earthquake vibrations originate in the surface of fracture; the surface from which they start has at first a very small area, which may quickly become very large, but at a rate not greater than the velocity of compressional elastic waves in the rock.

This extract is taken from Reid's statement of his elastic rebound theory (see §6.1), which has been a leading principle in earthquake seismology for several decades. In spite of his foresight, however, rupture-propagation effects were not observed until 1960 when the Chilean earthquake occurred.

## 5.1.1 *Phase comparison*

Benioff, Press & Smith (1961) applied a rupture-propagation model to the Chilean earthquake of 22 May 1960, and proved that the observed phase shift between vertical and horizontal components of the Earth's free oscillation can be well explained if rupture propagation is assumed to take place at the source.

In order to formulate the basic problem, let us first take a two-dimensional model. A force is applied to a point on a disc representing the Earth (fig. 5.1), exciting free oscillation in it. The surface displacements ($u_r$ and $u_\theta$ for the radial and tangential components, respectively) are given by:

$$u_r^{(n)} = - R(r) \cos n\theta \cos \omega t, \qquad (5.1)$$

$$u_\theta^{(n)} = \Theta(r) \sin n\theta \cos \omega t,$$

where $n$ is the mode number, and $R$ and $\Theta$ are the amplitude functions of the radial and tangential components respectively. Note that $u_r$ and $u_\theta$ in (5.1) have the same phase. This is physically understandable if we recall the phase relation in a standing wave. Let us assume symmetric radiation for the Rayleigh waves from the source, that is to say, two wave trains of equal period and of equal amplitude are travelling around the disc in opposite directions. The fundamental features of wave interference in this case may be found by considering a model of a semi-infinite medium in which two opposing trains of Rayleigh waves meet at a surface station.

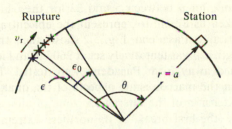

Fig. 5.1. Section of the disc model showing the parameters used in the analysis of radiation from a one-directional source propagating with a velocity $v_r$. The symbols $a$, $\varepsilon$ and $\varepsilon_0$ denote the radius of the disc, angular increment of the moving source, and angular amplitude of the source region respectively.

The waves of each train are polarized in such a manner that the particles of the medium (e.g. at its surface) move in a vertical plane parallel to the direction of wave propagation. The particle motion is elliptic and such that the vertical displacement (sinusoidal in time) leads the horizontal displacement by 90° for wave propagation in the positive direction, or lags it by 90° in the negative direction along the surface (see Bullen, 1953). Consequently, if two opposing trains of identical Rayleigh waves meet at a station, the vertical and horizontal components of the resultant ground motion will have phase differences of 0° and 180° as is the case in a standing wave.

If the source moves along the rim, however, the radiation cannot, in general, be symmetric. We consider here the case of one-direction source movement with a velocity $v_r$. Then, application of an appropriate delay to $u_r$ and $u_\theta$ in (5.1) yields

$$u_r^{(n)} = - R_n(a) \int_0^{\varepsilon_0} \cos\left[n(\theta - \varepsilon)\right] \cos\left\{\omega_n[t - (a\varepsilon/v_r)]\right\} \, d\varepsilon,$$

$$u_\theta^{(n)} = \Theta_n(a) \int_0^{\varepsilon_0} \sin\left[n(\theta - \varepsilon)\right] \cos\left\{\omega_n[t - (a\varepsilon/v_r)]\right\} \, d\varepsilon,$$

(5.2)

where $a$, $\varepsilon$, and $\varepsilon_0$ are the radius of the disc, angular increment of the moving source from its initial position, and angular amplitude of the source region, respectively (fig. 5.1). As a result of asymmetric radiation, the phase shift between the vertical and horizontal displacements at the station will vary in a complicated manner, between 0° and 180°. Using (5.2), we obtain the phase shift $\delta\phi_n$ as a function of parameters such as $n$, $\theta$, $\varepsilon_0$ and $v_r$ as

$$\delta\phi_n = (\pi/2) + \phi_r^{(n)} + \phi_\theta^{(n)},$$

(5.3)

where $\phi_r^{(n)}$ and $\phi_\theta^{(n)}$ denote the phase angles of $u_r$ and $u_\theta$, respectively.

Benioff, Press & Smith (1961) extended this model to three-dimensions and calculated $\delta\phi_n$ for $n$ between 5 and 33 for the Chilean earthquake, where $\theta$ was taken as 82.91° to represent the epicentral distance to the observation station at Pasadena. Fig. 5.2 shows their theoretical results. The rupture length $a\varepsilon_0$ was tentatively set at 960 km and four ruptures with velocities directed away from Pasadena were tested. This value seems reasonable when the macroseismic aspects of the quake are considered (north–south extension of the aftershock zone of almost 1000 km, with the epicentre, i.e. the first break, at its northern extremity).

Sets of line segments in fig. 5.2 represent the observations at Pasadena. Phase shifts are derived from linear strainmeters and gravimeters, which record the horizontal and vertical components of the free oscillation, respectively. The original plots, which show considerable scatter, are smoothed to get the lines shown in the figure. Comparing observation with theory, we may choose the rupture velocity to be 3 to 4 km s$^{-1}$.

The *phase-comparison method*, discussed above, has not been applied as frequently as the directivity-function method (see §5.1.2), because earthquakes as large as the 1960 event do not occur very often. Yet, it has proved an important contribution to earthquake seismology. The first successful recognition of rupture velocity strongly stimulated progress in this research field.

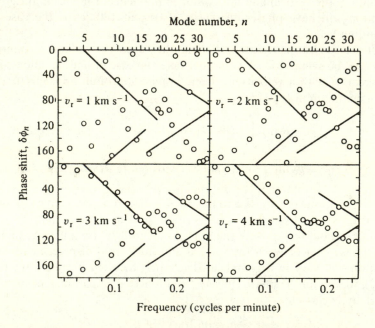

Fig. 5.2. Theoretical phase shifts between vertical and horizontal components for the Chilean earthquake (1960) for an assumed fault length of 960 km and several rupture velocities. (From Benioff, Press & Smith, 1961.)

### 5.1.2 *Directivity function*

The directivity-function method contrasts with the phase-comparison method in several respects. It compares the spectral amplitudes of surface waves leaving the source in opposite directions. As will be seen later, the effects of directional propagation of rupture are greatest in the period range equivalent to the process time at a fault-origin, i.e. 50–300 s for a magnitude-8 earthquake. The directivity-function method may also be applied to observations from smaller earthquakes provided that the surface waves that have circled the Earth can be accurately measured.

Let us follow Ben-Menahem (1961) and assume the geometry of a rectangular fault-origin shown in fig. 5.3. A fracture element along a line segment of length $D$, which is perpendicular to the fault strike, propagates in the plane $y = 0$ parallel to the $x$-axis for a distance $L$ with a uniform velocity $v_r$. Let us assume a teleseismic case and let $L/r$, $D/L$, and $\lambda/r$ all be much smaller than unity (where $r$ is the epicentral distance and $\lambda$ is the wavelength). Then, the *directivity function*, which is the ratio of spectral amplitude, $A$, of a surface wave leaving the source in the azimuth $\theta$ to the station to that in the opposite azimuth $\theta + \pi$, is given by

$$\text{Directivity} = \left| \frac{A(\theta)}{A(\theta + \pi)} \right| = \left| \frac{\left( \dfrac{c}{v_r} + \cos \theta \right) \sin \left[ \dfrac{\omega L}{2c} \left( \dfrac{c}{v_r} - \cos \theta \right) \right]}{\left( \dfrac{c}{v_r} - \cos \theta \right) \sin \left[ \dfrac{\omega L}{2c} \left( \dfrac{c}{v_r} + \cos \theta \right) \right]} \right|,$$

(5.4)

where $c$ is the phase velocity of the surface waves under study.

Båth (1974) explains the mathematical derivation of (5.4) in a simpler

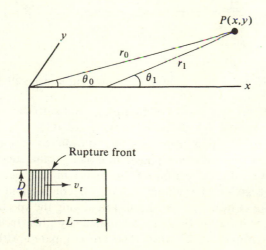

Fig. 5.3. Geometry of a rupture-propagating source with a station at the point $P$.

way than Ben-Menahem (1961). From a point $x$ on the fault (see fig. 5.4), the emitted wave arrives at the receiver with a time delay of

$$\tau_0 = (x/v_r) + (r - x \cos \theta)/c.$$

Therefore, the transfer function for the whole fault is

$$\frac{1}{L} \int_0^L \exp\left\{ i\omega \left[ t - \left( \frac{x}{v_r} + \frac{r - x \cos \theta}{c} \right) \right] \right\} dx = \frac{\sin X}{X} e^{-iX} e^{i\omega(t - r/c)}, \qquad (5.5)$$

where $X$ for the case of a vertical strike-slip fault is given by

$$X = \frac{\omega L}{2c} \left( \frac{c}{v_r} - \cos \theta \right) = \frac{\pi L}{\lambda} \left( \frac{c}{v_r} - \cos \theta \right), \qquad (5.6)$$

for wave propagation along the minor arc to the station. For the major arc path, on the other hand, $-\cos \theta$ in (5.6) should be replaced by $+\cos \theta$. Consequently, (5.4) is obtained for the amplitude ratio of the transfer functions for the minor and major arc seismic paths. We have assumed that *unilateral* faulting occurs, i.e. a fracture along a line segment $D$ (fig. 5.3) propagates only in one direction. If fracture propagation occurs in both directions, as in *bilateral* faulting, the directivity is more complex.

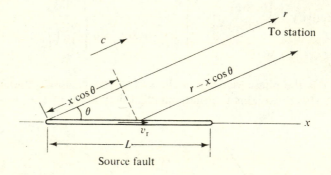

Fig. 5.4. Wave radiation from a propagating source. (After Båth, 1974.)

If a station records a series of multiple-circuit surface waves of the same mode, such as R1, R2, R3, ..., or G1, G2, G3, ..., we can obtain the directivity function from a single station by comparing spectral amplitudes of two successive phases in either group, such as R1 and R2 or G1 and G2 after correction for differences in propagation conditions. One advantage of this method is that all common factors, including the recording conditions, are cancelled by taking the spectral ratio.

The spectral ratios (R2/R1) calculated by this technique for the Chilean earthquake of 22 May 1960 are compared with the observed values in fig. 5.5.

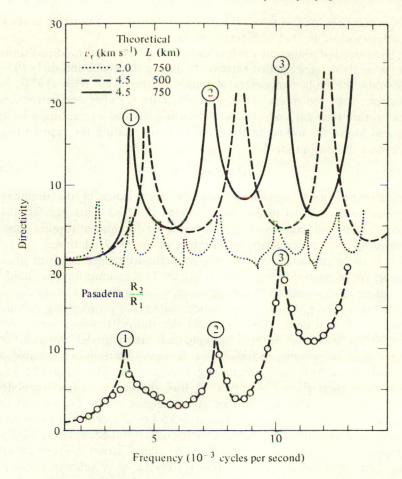

Fig. 5.5. A comparison of calculated spectral ratios (R2/R1) with observed spectral ratios (recorded at Pasadena) for the Chilean earthquake of 22 May 1960. (From Press, Ben-Menahem & Toksöz, 1961.)

The positions of the maxima and minima of the directivity function are critical when fitting the model to the observations. Taking the focal mechanism and other macroseismic data into consideration, Press, Ben-Menahem & Toksöz (1961) selected the set of parameters: $L = 750$ km and $r_r = 4.5$ km s$^{-1}$. Note that the theoretical curves for this model explain the observed curve for the Rayleigh waves well so far as the basic characteristics are concerned. This set of parameters is not precisely the same as the parameters used in the phase-comparison method discussed above. If we consider the typical errors of the two methods, however, we can see that there is approximate agreement between the two sets of parameters, proving that the Chilean earthquake was a large rupture about 1000 km long, in which fracture propagated from north to south at a velocity close to that of S-waves in the crust.

The directivity-function method described above, and modified forms of the method, were applied extensively during the 1960s and early 1970s to study the fault parameters of major earthquakes. Båth (1974), for example, collected much of the available data together and introduced the rupture time $L/v_r$, which is fault length $L$ divided by rupture velocity $v_r$, and found the following empirical formula relating the rupture time $L/v_r$ and the magnitude $M$:

$$\log (L/v_r) = 0.5 \, M - 1.9, \tag{5.7}$$

which shows good agreement with the dependence of the dominant spectral period, $T$, on magnitude discussed in §2.3.3. This suggests that longer dominant periods should be expected for longer rupture times, providing us with a realistic explanation of the $M$–$T$ relation.

It must be remarked that the above method assumes excellent phase coherence of seismic waves from an origin. This assumption may hold if a rupture is very smooth. If not, however, which is likely to be the case in an actual rupture, the phase coherence claimed for propagating ruptures does not in fact occur. To counteract this difficulty, seismologists synthesize surface waves using a moving-dislocation model and compare them with teleseismic records from as many stations as possible at various azimuths from the origin (e.g. Kanamori, 1970; see also §5.5.4 for synthetic seismograms). Appendix 1 lists rupture velocities calculated from these methods for a number of earthquake events. Notice that the rupture velocities are mostly close to 3–3.5 km s$^{-1}$, which is the typical shear wave velocity in the crust. This seems natural because the fracture propagation is controlled essentially by shear strain energy. Energy transport with velocity greater than the elastic wave velocity would be dissipative, as seen in supersonic shock waves. The evidence given above for the rupture-propagation theory is extremely interesting, for it supports the fracture mechanism at the seismic origin suggested by Reid (see §5.1). It is worth noting that as far as the present data are concerned, the ultimate fracture velocity is more likely to depend on the shear wave velocity than on the compressional wave velocity.

## 5.2    BODY WAVES FROM A FINITE MOVING SOURCE

### 5.2.1    *A moving dislocation model*

The force–dislocation equivalence (§4.3.2) shows that the displacement field generated by dislocation on a small element of the discontinuity surface is equivalent to that generated by a double-couple force applied at the surface element. The dynamical theory of dislocation, described below, is based on this equivalence (see for example Maruyama, 1963; Burridge & Knopoff, 1964).

A shear fracture, for example, is mathematically equivalent to a surface distribution of double-couple force systems over the entire fault surface, but the appropriate time delay should be taken into account to represent the fracture propagation. In other words, the seismic displacement field due to this model is obtained in principle by introducing the correct terms for the time delay into the displacements given in (3.12). We recall that the far-field displacement is represented by a time function proportional to the time derivative of the stress change (and consequently the offset) at the surface element.

Let us take the model of fracture propagation introduced in §5.1.2 (fig. 5.3). We assume that the offset about a line segment varies in a step time function, and that the line segment moves with a constant velocity $v_r$ in a direction parallel to the $x$-axis. Then the P- and S-waves far away from the source will appear as rectangular (box-car type) pulses, and their time width will be given by

$$\tau_0 = (L/v_r) - (L/c) \cos \theta, \tag{5.8}$$

where $c$ is the velocity of the P-waves ($v_P$ or $\alpha$) or S-waves ($v_S$ or $\beta$), and $\theta$ is the angle between the direction of fracture propagation and the seismic ray to the station.

Hirasawa & Stauder (1965) considered a vertical strike-slip fault with unilateral fracture propagation, and derived the radiation patterns of P- and S-waves. Fig. 5.6 shows the amplitudes of P- (radial component) and S- (transverse component) waves in the $xy$-plane, for the case of unilateral fracture propagation in the positive $x$-direction. The effect of fracture propagation is evident, especially when $v_r/v_S = 0.9$, if these diagrams are compared with those for a point source. It should be noted, however, that the product $A\tau_0$ (where $A$ is the amplitude of the seismic pulse) is approximately constant for different azimuths. Consequently, for the positive direction of the $x$-axis where the amplitude is very large, the pulse width is expected to be very narrow.

### 5.2.2 Stopping phase

The first term in (5.8) represents the time interval between the initiation and termination of the fracture, whereas the second term represents the difference between the two travel times to the station, one from the point of initiation and another from that of termination. Therefore we may rewrite (5.8) as

$$\tau_0 = L/v_r + T(\Delta', h') - T(\Delta, h),$$
$$\Delta' = \Delta + \delta\Delta, \quad h' = h + \delta h, \tag{5.9}$$

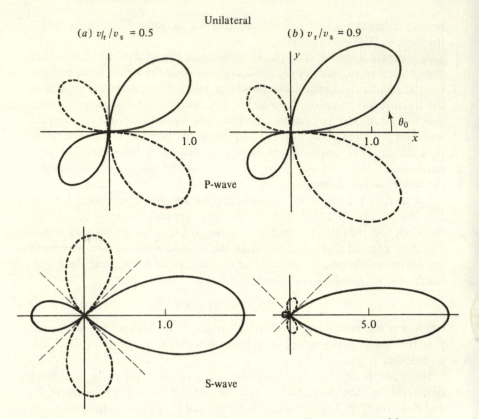

Unilateral
(a) $v_r/v_s = 0.5$ (b) $v_r/v_s = 0.9$

P-wave

S-wave

Fig. 5.6. Radiation patterns of the P- and S-waves from sources with rupture propagation for: (a) $v_r/v_s = 0.5$; (b) $v_r/v_s = 0.9$, where $v_r$ and $v_s$ denote the rupture and shear wave velocities respectively. Solid lines indicate positive, and dashed lines negative values. (From Hirasawa & Stauder, 1965.)

where $T$ is the travel time, $\Delta$ and $h$ denote the epicentral distance and depth for the point of initiation and $\Delta'$ and $h'$ are the epicentral distance and depth for the point of termination. Consequently, the travel-time plot of the pulses will appear as if two independent shocks occurred at adjacent points with a time interval equal to $L/v_r$. In fact, Savage & Mansinha (1963) conducted a laboratory experiment on tensile crack propagation in a glass plate and recognized a special phase after P, which is attributed to the termination of rupture. This phase is therefore called the *stopping phase*. It is characterized by a polarity opposite to that of the initial P-pulse, and reaches the station after P with a time delay $\tau_0$ given by (5.8) or (5.9).

Hirasawa (1965) identified this phase on a group of seismograms of the Niigata earthquake, Honshu, Japan (16 June 1964, $M = 7.5$) and estimated the source parameters using the following argument. If $\delta\Delta$ and $\delta h$

in (5.9) are small, we obtain:

$$\tau_0 = \frac{L}{v_r} - \frac{a}{a-h} L \sin i_0 \cos(\theta_N - \theta_0)\left(\frac{\partial \Delta}{\partial T}\right)^{-1} + L \cos i_0 \left(\frac{\partial h}{\partial T}\right)^{-1}, \quad (5.10)$$

where,

> $a$ is the radius of the earth,
> $i_0$ is the angle of the direction of fracture propagation measured from the downward vertical,
> $\theta_0$ is the azimuth of the direction of fracture propagation measured from the north,
> $\theta_N$ is the azimuth of the ray path to a station measured from the north.

In the present case of shallow focus and of horizontal propagation of fracture (perhaps symmetric bilateral – see §5.3), (5.10) reduces to a much simpler form since $\sin i_0 \approx 1$, $\cos i_0 \approx 0$ and $a/(a-h) \approx 1$. In fig. 5.7, $\tau_0$ read from teleseismic records is plotted against $(\partial T/\partial \Delta)\cos(\theta_N - \theta_0)$. Equation (5.10) reduces to $\tau_0 = (L/v_r) - L[(\partial T/\partial \Delta) \cos (\theta_N - \theta_0)]$ so that $L$ and $v_r$ can be determined by fitting a straight line to the data in fig. 5.7 by the method of least-squares. In the case of bilateral faulting, two straight lines must be fitted as shown in the diagram ($\theta_0$ is N 10°E in this case). In this way, Hirasawa obtained a fault length of $40 \pm 10$ km and a rupture velocity of $2.0 \pm 0.4$ km s$^{-1}$, for the northeast branch of the bilateral fault.

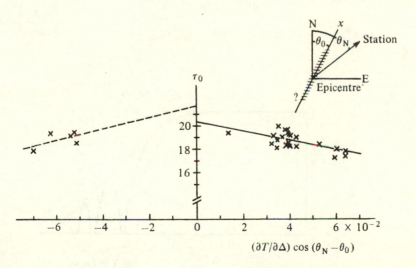

Fig. 5.7. Arrival time differences, $\tau_0$ (in seconds), between the $P_3$- and $P_1$-phases for the Niigata earthquake (1964). The geographical relation of a station to the fault source is illustrated in the top right of the figure. (From Hirasawa, 1965.)

### 5.2.3   *Corner frequency*

In §5.2.1 we learned that a finite moving source radiates P- and S-pulses of rectangular shape with width $\tau_0$. Suppose the pulse height of a wave is $A_0$, its spectrum is given by:

$$U(\omega) = A_0\tau_0(\sin X)/X,\qquad (5.11)$$

with

$$X = (\omega\tau_0/2).\qquad (5.12)$$

Fig. 5.8 illustrates the spectrum on a log–log plot.

The basic structure of this spectrum is characterized by a plateau for low frequencies and a downward sloping envelope for high frequencies. The two trends, given by $\omega^{-1}$ and $\omega^{-2}$ and represented by broken lines on the figure, intersect at $X = 1$. The frequency corresponding to this value of $X$ is called the *corner frequency* and is generally denoted by $\omega_0$. From (5.8), we obtain for $\omega = \omega_0$:

$$L = 2c/[\omega_0(c/v_{\mathrm{r}} - \cos\,\theta)].\qquad (5.13)$$

That is to say, if we fit the theoretical curve to an observed spectrum corrected for the propagation and observation conditions, we can estimate $L$ from the corner frequency, as a function of $c$, $v_{\mathrm{r}}$, and $\theta$.

Brune (1970) assumed the following source time function for displacement (see §5.4.1)

$$u(t) = (\sigma/\mu)\beta\tau(1 - \mathrm{e}^{-t/\tau}),\qquad (5.14)$$

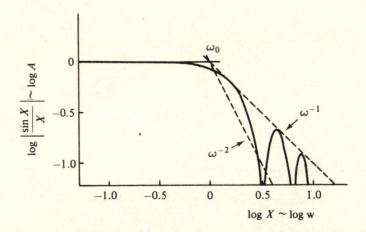

Fig. 5.8. Idealized body-wave spectrum from a finite moving source. (From Båth, 1974.)

where $\sigma$ is the effective stress (cf. §6.2.3), $\mu$ the rigidity, $\beta$ the shear wave velocity and $\tau$ the time constant. Notice that his model does not consider explicitly the effects of fracture propagation. Implicity, however, the effect may be included in $\tau$, which is of the order of the dimension of the fault divided by the shear wave velocity. The spectrum of displacement at the source becomes, in this case,

$$U(\omega) = (\sigma\beta/\mu)/\omega(\omega^2 + \tau^{-2})^{1/2}, \tag{5.15}$$

and the far-field spectrum (shear wave) is given, taking its rms (root-mean-square) average, by

$$\overline{u(\omega)} = \overline{R_{\theta\phi}}(\sigma\beta/\mu)(a/r)F(\varepsilon)/(\omega^2 + \alpha^2)^{1/2} \tag{5.16}$$

where $\overline{R_{\theta\phi}}$ is the rms average of the radiation pattern, $a$ is the radius of an equivalent circular dislocation surface, $r$ is the distance from the source, $\varepsilon$ is the fraction of stress drop, $\alpha = 2.21\ \beta/a$ and

$$F(\varepsilon) = \{(2 - 2\varepsilon)[1 - \cos\ (1.21\varepsilon\omega/\alpha)] + \varepsilon^2\}^{1/2}. \tag{5.17}$$

On the basis of this model, Brune (1970) and later Hanks & Thatcher (1972) (see fig. 5.9) discussed the basic spectral parameters in terms of physical parameters at a source (see §5.4.1). As seen in fig. 5.9, the far-field displacement spectrum may be described by three basic elements in the simplest case, i.e., the amplitude at zero frequency, the corner frequency and the slope at frequencies above the corner frequency (log–log plot).

Fig. 5.10 shows an example of Brune's model (see also §5.4.1) applied to the Borrego Mountain, California, earthquake of 9 April 1968. Success-

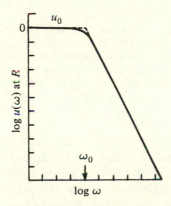

Fig. 5.9. The far-field shear displacement spectrum of Brune (1970). Vertical and horizontal axes are divided into arbitrary logarithmic units (the fraction of the stress drop, $\varepsilon$, is assumed to be 1). (After Hanks & Thatcher, 1972).

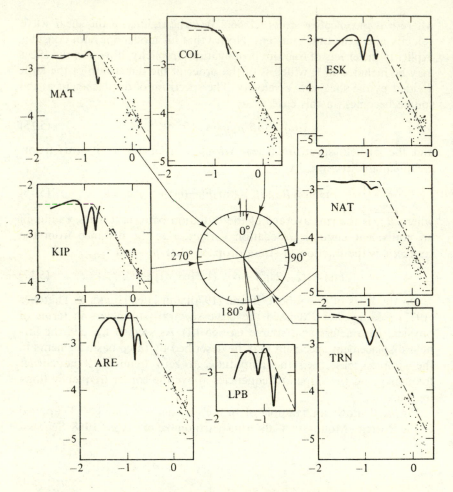

Fig. 5.10. P-wave spectra for the California earthquake, 9 April 1968, at different azimuths from the fault trace, N48 W. Solid lines are spectral data from long-period instruments; points are spectral data from short-period instruments. Vertical scales are log amplitude spectral density (where spectral densities are measured in cm s) and horizontal scales are log frequency (where frequencies are measured in cycles per second). The abbreviations for station names refer to the WWSSN code, and the station names and locations are given in appendix 3. (From Hanks & Wyss, 1972.)

ful comparison of a theory with observed spectra will also be seen in §5.5.2 in the case of the Imperial Valley (or El Centro) earthquake of 18 May 1940. Reading accuracy of a corner frequency depends greatly on the accuracy of fitting the two envelopes to observations, which is sometimes difficult, especially on the plateau. Nevertheless, the corner-frequency method has been applied widely even to microearthquakes, to which the directivity method does not apply (e.g. Douglas & Ryall, 1972; Ishida, 1974).

As this example demonstrates, spectral analyses of seismic waves provide us with powerful tools for studying source mechanisms. Further discussion will be developed in the following sections from various points of view (seismic moment in §5.3, source time function in §5.4, and further physical parameters in §5.5 and in chapter 6).

## 5.3 SEISMIC MOMENT

### 5.3.1 *The use of long-period surface waves*

Analyses of the static and dynamic fields of an event give two different ways to estimate the source parameters. As discussed in §4.3.2, the magnitude of a double-couple source may be represented by the moment of one component couple, and is given by $M_o = \mu U A$. The previous section (§5.2.3) introduced an alternative way to determine the source moment using the spectral amplitude (or density) at zero-frequency inferred from observed wave-forms. It is a theoretically interesting technique, but is often difficult to implement in practice since a long-range extrapolation of relatively high frequencies from body-wave spectra may often be erroneous.

Use of long-period surface waves is advantageous in this respect. Aki (1966*a, b*) analysed the spectra of G-waves in the Niigata earthquake (16 June 1964, $M = 7.5$). This earthquake is attributed to a reverse fault striking approximately N20°E, judging from P- and S-wave data (cf. §5.2.2) as well as from macroseismic properties (§5.3.2). In fact, a radiation pattern of the long-period spectral component ($T = 200$ s) of G2 waves, which was examined in the course of the study, confirmed the above model.

Fig. 5.11 summarizes the results of Aki's analyses. He first obtained the mean amplitude of spectral density of G2 waves equalized to a lapse time of 7000 s and to epicentral distance of 90° in the loop directions (azimuth 0 to $-30°$ and 150 to 180°). Then he introduced corrections for dissipation, instrumental conditions, and propagation along the spherical surface of the Earth. Consequently, the figure illustrates the mean spectral density of G2 in the loop direction, equalized to a distance of 10 000 km on a non-dissipative flat Earth model.

The smooth curves (1), (2) and (3) in fig. 5.11 represent theoretical far-field Love-wave displacement (spectral density), calculated using Haskell's (1964*a*) theory. To draw them, Aki (1966*b*) assumed a step time function for the variation of the moment of component couple in the

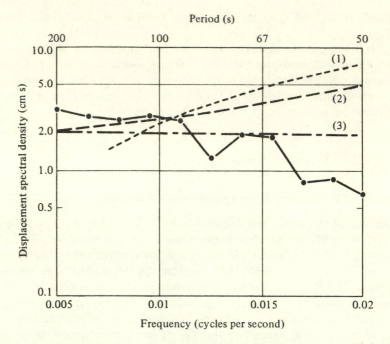

Fig. 5.11. Comparison of the observed (solid line) displacement spectral density in the loop direction (equalized to a distance of 10 000 km on a non-dissipative flat-Earth model) with the theoretical predictions (dashed curves (1), (2), (3)). The theoretical curve (1) corresponds to an Earth model with a single-layer crust overlying a uniform mantle. The curve (2) corresponds to a Gutenberg (multi-layer) mantle model. The curve (3) is the curve (2) corrected for the effect of finiteness of the earthquake source. In all theoretical curves, the moment of the source couple is assumed to be $3 \times 10^{27}$ dyn cm. (After Aki, 1966*b*.)

double-couple force at the origin, such that

$$M_o = 0 \qquad \text{for } t < 0,$$

$$M_o = 3 \times 10^{27} \text{ dyn cm} \quad \text{for } t \geqq 0.$$

(5.18)

Haskell's theory enables us to study Love waves from a point source, in a surface layer over a semi-infinite medium (case 1), with unit component moment varying in time as $e^{i\omega t}$. The spectrum of waves using a step-wise varying moment may be synthesized from his formula. Curve (1) represents the case of a single surface layer, whereas (2) and (3) represent modified models as explained in the caption. Briefly, the amplitude of seismic moment at the origin ($3 \times 10^{27}$ dyn cm) explains the observations at the frequencies of about $\omega = 0.01$ cycles per second fairly well, provided scattering of the gradients of the curves are disregarded. A more straightforward determination of seismic moment, by comparing observed surface waves with theoretical records synthesized from a source model to be tested, has recently become popular – see §5.5.

### 5.3.2  Implications of the seismic moment

*Average dislocation.* The force–dislocation equivalence, which is valid in the static and dynamic cases (§4.3.2 and §5.2.1), has yielded a formula relating the seismic moment to the average dislocation across a fault origin.

Aki (1966*b*) took a rectangular fault model in which the fault plane undergoes unidirectional uniform offset with a step time function, and related the average dislocation, $U$, to the moment, $M_o$, obtained from seismic waves using

$$M_o = \mu U A, \tag{5.19}$$

which corresponds to (4.20) for the static field (where $\mu$ is rigidity and $A$ is fault area).

In the case of the Niigata earthquake, $M_o = 3 \times 10^{27}$ dyn cm from surface wave observations (§5.3.1). Also, we may take the fault area as 100 km (length) $\times$ 20 km (width), and the rigidity as $3.7 \times 10^{11}$ dyn cm$^{-2}$ from the crustal parameters used in the previous analyses. Then we obtain:

$$U = 400 \text{ cm}. \tag{5.20}$$

This agrees with the field evidence for the fault which indicates that the maximum sea-bottom uplift and subsidence exceed 5 m and 4 m respectively, as revealed by repeated echo-sounding surveys (Mogi, Kawamura & Iwabuchi, 1964.)

*Energy released and stress-drop.* Aki compared the above fault to the Starr model and referred to a formula given by Starr (1928):

$$E = \frac{\pi}{4} \sigma^2 D^2 \frac{\lambda + 2\mu}{\mu(\lambda + \mu)} L, \tag{5.21}$$

for the energy released by formation of a long strip of crack (slip perpendicular to the direction of the strip) in an infinite body under uniform shear stress $\sigma$ ($L$ is the length, $D$ is the half-width of the strip crack, and $\lambda$ and $\mu$ are Lamé's constants). He obtained:

$$E = \frac{8}{3\pi} U^2 \mu L, \tag{5.22}$$

where he assumed that the fault plane is stress free after slipping, that

$\lambda = \mu$ and

$$U = (\pi/4)U_m. \tag{5.23}$$

The symbol $U_m$ denotes the maximum relative displacement (i.e. offset) and is given by Starr as:

$$U_m = \sigma D(\lambda + 2\mu)/[\mu(\lambda + \mu)]. \tag{5.24}$$

Assigning the appropriate values to the parameters, we obtain

$$E = 5.0 \times 10^{23} \text{ erg}, \tag{5.25}$$

for the Niigata earthquake. This is reasonably larger than the seismic energy, $1.1 \times 10^{23}$ erg, known from the magnitude ($M = 7.5$).

Equation (5.24) is used to estimate the pre-existing stress $\sigma$ as

$$\sigma = \tfrac{2}{3} \mu U_m/D = \frac{8}{3\pi} \mu U/D, \tag{5.26}$$

assuming $\lambda = \mu$. Taking $D$ as 10 km, we obtain a stress

$$\sigma = 126 \text{ bar}, \tag{5.27}$$

and the corresponding strain ($\varepsilon = \sigma/\mu$)

$$\varepsilon = 3.4 \times 10^{-4}.$$

This value of strain agrees well with the near-field evidence (land tilting of about 1 minute of arc, i.e. $3 \times 10^{-4}$ rad, was observed at a small island, Awashima, located several kilometres from the epicentre by Nakamura, Kasahara & Matsuda, 1964). Also, this value of strain is consistent with the ultimate strain of the crust discussed in §2.3.4.

From (5.21) and (5.24), we obtain a new expression for $E$:

$$E = \tfrac{1}{2}\sigma U A. \tag{5.28}$$

Combination of (5.28) and (5.19) yields

$$\frac{E}{M_0} = \frac{\sigma}{2\mu}, \tag{5.29}$$

in which the parameters specifying the fault's behaviour, such as the area $A$ and offset $U$, cancel. If the Starr (1928) model applies to the source, therefore, we may estimate the pre-existing stress without these parameters. This is extremely advantageous for studying deep-focus earthquakes and microearthquakes, for which near-field data are unavailable. In practice, however, the efficiency of seismic radiation must be known correctly in order that we may reduce the seismic energy to obtain the total energy release, $E$, in the event.

The above discussion has demonstrated the usefulness of seismic moment in the study of source mechanics. The source parameters, which are derived from seismic wave data via a seismic moment, are in good agreement with the information from the macroseismic field evidence, in spite of the many assumptions and simplifications made in the analyses. This seems to suggest that the seismic moment technique is quite reliable, and more fundamentally, that the fault-origin model is reasonable.

### 5.4   SOURCE TIME FUNCTION

### 5.4.1   *Kinematic considerations*

The foregoing discussions have demonstrated that useful information about the fault's behaviour is contained in the seismic pulse shape. In order to express wave radiation in terms of the source parameters, slip or stress change on the fault surface are often assumed to have a step time function (see §5.2 and §5.3 for examples). This is advantageous in that it leads to mathematical simplicity. But physically, it is not plausible, as such an abrupt change would require a singularity to occur in other source parameters. This suggests that the source time function must be modified in such a manner as to have a finite initial gradient. Two alternative models have been proposed for this purpose, and these are introduced below.

*Haskell model.* Let us consider a moving dislocation model with the geometry shown in fig. 5.3. If we express the displacement, $u(t)$, on the fault surface in the form

$$u(t) = u_\infty H(t - x/v_r), \tag{5.30}$$

with $u_\infty$ and $H(t)$ denoting the final displacement (half the fault slip $U$ in the simplest case), and the Heaviside *unit step function* ($H(t) = 0$, $t < 0$; $H(t) = 1$, $t \geq 0$), respectively, then we have the propagating *Heaviside dislocation model* proposed by Knopoff & Gilbert (1959). Haskell (1964b) replaced (5.30) by the following expression

$$u(t) = u_\infty G(t - x/v_r), \tag{5.31}$$

to construct a *ramp dislocation model*, where $G(t)$ represents a *ramp function*. This function is zero at $t < 0$ and increases linearly with time until it reaches 1 at $t = T$, which is called the *rise time*, or the characteristic time (for asymptotic curves, the time to reach 90% of the final level of the curve is sometimes called the rise time – see §5.4.2). The function in (5.31) represents, therefore, a process in which slips of this type occur progressively along a fault with a velocity $v_r$, resulting in a uniform slip over the whole fault surface.

Fig. 5.12 compares the source time functions of the two models together with the source time function for the next model, by Brune.

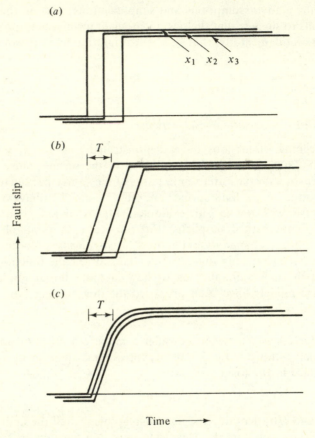

Fig. 5.12. Source time functions in three source models; (a) after Knopoff & Gilbert (1959); (b) after Haskell (1964b); and (c) after Brune (1970).

*Brune model.* To illustrate the model of Brune (1970) let us assume that a tangential stress step is applied to the interior of a dislocation surface causing the fault block on one side to move in one direction, and the block on the other side, in the opposite direction. The step is assumed to apply instantaneously over the fault surface, that is to say, the fault propagation effects are neglected, for simplicity. Also, it is assumed that the fault surface reflects elastic waves totally during rupture, i.e. the elastic events on the two sides of the fault are isolated from one another by the fault surface.

The stress step sends a pure shear stress wave propagating per-pendicularly to the fault surface, so the initial time function of the pulse is given as

$$\sigma_0(y, t) = \sigma H(t - y/\beta), \tag{5.32}$$

where $\beta$ is the shear wave velocity and $H(t)$, the Heaviside function. $\sigma$, on the right-hand side of the equation, is the effective stress ($\sigma_{\text{eff}}$ in Brune, 1970), but we will abbreviate it here, for simplicity.

If $\mu(\partial u/\partial y) = -\sigma_0(y, t) = -\sigma H(t - y/\beta)$, and $u = 0$ when $t = y/\beta$, then

$$u = (\sigma/\mu)\beta(t - y/\beta)H(t - y/\beta)$$

and

$$\dot{u} = (\sigma/\mu)\beta H(t - y/\beta). \tag{5.33}$$

At a point near the fault, therefore, displacement increases linearly in time as the stress pulse propagates away from the fault. Then it levels off as the finiteness of the fault becomes felt at the observation point. This rise time depends on the distance of the observation point from the fault's extremity, and is longest at the centre of the fault. The following discussion will be concerned with this central point, unless otherwise stated.

The initial particle velocity at $y = 0$, $t = 0$ is given by

$$\dot{u}_0 = (\sigma/\mu)\beta. \tag{5.34}$$

If $\sigma = 10^8$ dyn cm$^{-2}$, $\mu = 3 \times 10^{11}$ dyn cm$^{-2}$, and $\beta = 3 \times 10^5$ cm s$^{-1}$, which are likely values in a shallow crust, the equation gives $\dot{u}_0 = 100$ cm s$^{-1}$. Strong-motion seismographs have recorded the near-source particle velocity in several earthquakes (Brune, 1970). For example, a maximum value of 76 cm s$^{-1}$ was observed at a site very near the fault trace in the Parkfield earthquake, California (28 June 1966, $M_S = 6.4$). Judging from these data, 100 cm s$^{-1}$ derived above may be a good estimate of the motion in a fault. In fact, the rise time inferred from seismic data also confirms this idea (Kanamori, 1973; cf. §5.4.3). This also suggests that $\sigma$ in large earthquakes is of the order of $10^8$ dyn cm$^{-2}$, again in good agreement with observations (see §§4.2.3, 5.3.2 and 5.5.4).

The particle velocity given by (5.34) will begin to decrease to zero when the effects of fault finiteness reach the observation point from the fault's extremities. Brune (1970) introduced a *time constant*, $\tau$, equivalent to the travel time of this signal, $\tau \sim a/\beta$, where $a$ is the equivalent radius of the fault surface, and replaced (5.33) by

$$u(y = 0, t) = (\sigma/\mu)\beta\tau(1 - e^{-t/\tau}), \tag{5.35}$$

$$\dot{u}(y = 0, t) = (\sigma/\mu)\beta e^{-t/\tau}. \tag{5.36}$$

The Fourier transform of (5.35) has been given by (5.16).

*Models in earthquake seismology.* Fig. 5.12 compares the source time functions of the step-function, ramp-function (Haskell) and the Brune models. The Haskell and Brune models are not equally applicable to all practical problems. The Haskell model describes the effects of rupture propagation well, but the use of a ramp-type source function is rather arbitrary and makes further analysis difficult in comparison with the case of an analytical function. The Brune model, on the other hand, employs an exponential function, so that mechanical considerations in the vicinity of a fault may be dealt with relatively easily (see also chapter 6). However, it does not consider explicitly rupture propagation with a finite velocity; in other words, it does not apply well to far-field problems in which the effect of rupture propagation is of special concern. To summarize, the Haskell and Brune models are complementary in application, and the most appropriate model must be chosen depending on whether far-field or near-field problems, respectively, are to be studied.

Simplicity and precision are always two competing elements in modelling. In the oversimplified models discussed above, several critical problems have been disregarded, in spite of their physical importance. One is that of singularities at the fault edge. This is especially serious in view of the previous assumption of uniform rupture velocity and uniform dislocation amplitude extending to the extremity of a fault. These idealized conditions are not physically realistic. A future advanced model, therefore, should take this, as well as the many other details of faulting, such as anelastic properties of the medium and irregular fracture propagation, into account.

Computer simulation provides us with a powerful tool for this purpose. Burridge & Knopoff (1967) proposed a one-dimensional model, which simulates a fault block by a group of discrete mechanical elements (associated with mass, elasticity, friction and viscosity), to explain repeated fault slips under stress. Further work by several researchers has led to a model of two-dimensional faulting associated with realistic frictional laws on the fault surface (§6.2.3).

### 5.4.2    *Characteristic times for the faulting process*

It has been suggested that a faulting process is basically described by two characteristic times. The first is related to rupture propagation along the fault with a velocity of about 3 km s$^{-1}$. The second is related to the particle motion to complete offset at any point on the fault, with a velocity estimated to be of the order of 100 cm s$^{-1}$. Suppose a fault is represented by a slip (half amplitude) of 2 m for an extension of 60 km, which is typical of an $M = 7.5$ earthquake (table 5.1), then the characteristic times associated with the first and second processes are estimated as 20 s and 2 s, respectively. A natural question arises about the manner in which the wave-form, or its spectrum, reflects these processes.

Table 5.1. *Average fault parameters for various magnitudes. (From Hasegawa, 1974.)*

| Magnitude $M_S$ | $M_0$ (dyn cm) | $L$ (km) | $D$ (km) | $U$ (cm) |
|---|---|---|---|---|
| 5.5 | $3 \times 10^{24}$ | 6 | 5 | 30 |
| 6.0 | $1.4 \times 10^{25}$ | 10 | 8 | 53 |
| 6.5 | $5 \times 10^{25}$ | 18 | 11 | 75 |
| 7.0 | $2 \times 10^{26}$ | 35 | 15 | 120 |
| 7.5 | $4 \times 10^{27}$ | 60 | 50 | 400 |

The far-field displacement spectrum, which Savage (1972) constructed on the basis of Haskell's theory (1964b), seems to answer this question to a certain extent. For the Haskell model discussed in §5.4.1, the far-field radiation is given in terms of particle velocity as

$$d\mathbf{u}/dt = \mathbf{R}_\alpha(\theta, \phi, r)\mu D I_\alpha + \mathbf{R}_\beta(\theta, \phi, r)\mu D I_\beta, \tag{5.37}$$

with

$$I_c = U \int_0^L \left[ \frac{d^2}{dt^2} G\left(t - \frac{r}{c} - \frac{\xi}{c}\left(\frac{c}{v_r} - \cos\theta\right)\right) \right] d\xi, \tag{5.38}$$

where $\theta$, $\phi$ and $r$ are the spherical polar coordinates shown in fig. 5.13, $\xi$ is the $x_1$-coordinate of the point of integration in the fault, $\mu$ is the rigidity of the medium, $\mathbf{R}_c$ is the radiation field from an appropriate double-couple source, $U$ is the final offset, and $c$ stands for either $\alpha$ or $\beta$, which denote the P- and S-wave velocities respectively. The first term in (5.37) represents the P-wave radiation and the second term the S-wave radiation. The factors $\mathbf{R}_c$ are given for longitudinal shear faulting (slip parallel to $x_1$) by

$$\mathbf{R}_\alpha = \mathbf{e}_r \sin 2\theta \sin \phi/(4\pi\alpha^3 \rho r),$$

$$\mathbf{R}_\beta = (\mathbf{e}_\theta \cos 2\theta \sin \phi + \mathbf{e}_\phi \cos \theta \cos \phi)/(4\pi\beta^3 \rho r), \tag{5.39}$$

where $\mathbf{e}_r$, $\mathbf{e}_\theta$, and $\mathbf{e}_\phi$ are unit vectors in the $r$, $\theta$, and $\phi$ directions and $\rho$ is the material density. Equations (5.37) and (5.38) give the integral seismic disturbances due to the moving rupture, which is represented by the propagating line segment (fig. 5.3).

Performing the integration in (5.38) yields:

$$I_c = (UL/\tau_0)\left[ \frac{d}{dt} G\left(t - \frac{r}{c}\right) - \frac{d}{dt} G\left(t - \frac{r}{c} - \tau_0\right) \right], \tag{5.40}$$

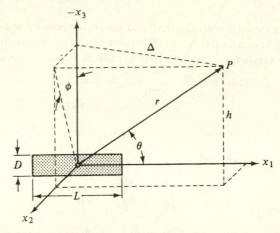

Fig. 5.13. Fault surface (shaded) and the coordinate system used in the study of focal process time. $P$ is the point of observation.

where

$$\tau_0 = (L/c)[(c/v_r) - \cos\,\theta]. \tag{5.41}$$

Hence, we can integrate (5.37) and obtain the far-field displacement $\mathbf{u}_c$ for each wave (P or S; suffix $c$ stands for either $\alpha$ or $\beta$) as

$$\mathbf{u}_c = \mathbf{R}_c(\theta,\,\phi,\,r)(\mu DLU/\tau_0)[G(t-(r/c)) - G(t-(r/c)-\tau_0)], \tag{5.42}$$

and obtain the amplitude of its Fourier transform as

$$|\mathbf{u}_c| = \mathbf{R}_c(\theta,\,\phi,\,r)M_0\omega|\hat{G}||F_1(\omega,\,\tau_0)|, \tag{5.43}$$

where $|\hat{G}|$ denotes the Fourier transform (amplitude) of $G(t)$,

$$F_1(\omega,\,\tau_0) = \sin(\omega\tau_0/2)/(\omega\tau_0/2), \tag{5.44}$$

and $M_0 = \mu ULD$.

Equation (5.42) can be used for bilateral rupture where slip propagates a distance $L_0$ in the $+x_1$ direction and a distance $L_\pi$ in the $-x_1$ direction. In this case, $|F_1(\omega,\,\tau_0)|$ in (5.43) should be replaced by $|F_2(\omega,\,\tau_0,\,\tau_\pi)|$, expressed as

$$F_2(\omega,\,\tau_0,\,\tau_\pi) = \{[L_0 F_1(\omega,\,\tau_0)]^2 + [L_\pi F_1(\omega,\,\tau_\pi)]^2$$
$$+ 2L_0 L_\pi F_1(\omega,\,\tau_0)F_1(\omega,\,\tau_\pi)\,\cos[\omega(\tau_0 - \tau_\pi)/2]\}^{1/2}/(L_0 + L_\pi), \tag{5.45}$$

where

$$\tau_0 = (L_0/c)[(c/v_r) - \cos\,\theta],$$
$$\tau_\pi = (L_\pi/c)[(c/v_r) + \cos\,\theta]$$

and now

$$M_0 = \mu U(L_0 + L_\pi)D.$$

For further discussion of (5.43) let us assume the following form for $G(t)$

$$G(t) = 0 \qquad \text{for } t < 0,$$

$$G(t) = 1 - e^{-t/\tau} \quad \text{for } t > 0.$$
(5.46)

Writing the Fourier transform of $G(t)$ as $|\hat{G}| = (1/\omega)/(1 + \omega^2\tau^2)^{1/2}$, the following expression is obtained for the spectral amplitude

$$|\mathbf{u}_c| = \frac{\mathbf{R}_c(\theta, \phi, r)M_0F_2(\omega, \tau_0, \tau_\pi)}{(1 + \omega^2\tau^2)^{1/2}}.$$
(5.47)

Fig. 5.14 illustrates the spectra of the far-field displacement signal, where parameters are assumed as follows: $L_0 = L_\pi (= L/2)$, $\theta = 60°$ and $v_r/\beta = 0.9$. The upper and lower pairs of curves correspond to $2\tau\beta/L = 0.005$ and $0.5$, respectively. The rise time at a point on the fault (i.e. the time to reach 90% of the final slip) is given by (5.46) as $2.3\tau$, which we may suppose to be roughly equal to $D/2v_r$. Since $v_r/\beta = 0.9$, we obtain

$$2\tau\beta/L \approx D/(2L).$$
(5.48)

That is to say, the upper and lower pairs of curves apply to fault surfaces having length to width ratios (aspect ratios) of $L/D = 100$ and $1.0$, respectively.

Notice that the form of these spectra is generally characterized by two basic trends, a flat asymptote (gradient 0) for lower frequencies and a slope of gradient $-2$ for higher frequencies which intersect at an intermediate frequency. On closer inspection, an intermediate slope of gradient $-1$ can be recognized, which is more noticeable in the upper pair, indicating the two corner frequencies $\omega_1$ and $\omega_2$ of the spectra.

The two corner frequencies $\omega_1$ and $\omega_2$ are respectively controlled by the factors $F_2(\omega, \tau_0, \tau_\pi)$ and $1/(1 + \omega^2\tau^2)^{1/2}$ in (5.47). Since $\tau_0 = \tau_\pi = L_0/v_r = L_\pi/v_r$ (if $\theta = 90°$), $2.3\tau \approx D/2v_r$ and $v_r = 0.9\beta$, we obtain the following relations:

$$\omega_1 \propto \beta/L,$$

$$\omega_2 \propto \beta/D.$$
(5.49)

In other words, $\omega_1$ and $\omega_2$ are related to the reciprocal of rupture-propagation time and of rise time, respectively. In fact, $\omega_1$ in fig. 5.14 is approximately $2\beta/L$ both in the upper and lower pairs, whereas $\omega_2/(\beta/L)$ is approximately 200 in the upper, but only 2 or 3 in the lower.

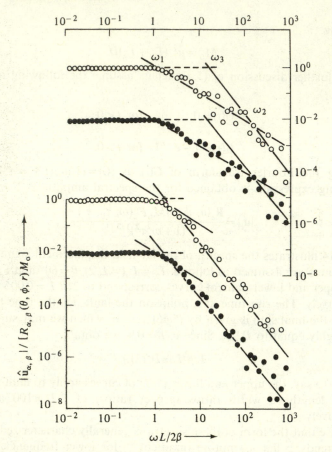

Fig. 5.14. The Fourier spectra of the far-field displacement signal calculated from the Haskell model (generalized to bilateral rupture) with $L_0 = L_\pi$, $\theta = 60°$ and $v_r/\beta = 0.9$. The ordinate for the P-wave spectra (solid circles) has been divided by 100 to separate the P- and S-wave spectra. In the upper pair of spectra, $2\tau\beta/(L_0 + L_\pi) = 0.005$ and in the lower pair, $2\tau\beta/c L_0 + L_\pi) = 0.5$; these values are appropriate to fault surfaces having length-to-width ratios, $L/D$, of 100 and 1.0 respectively. (From Savage, 1972.)

To summarize, a far-field spectrum of P- or S-waves is generally characterized by two corner frequencies, as a result of finite fault size, finite rupture velocity and finite slip velocity. Separation of the two frequencies depends on the ratio of the two characteristic process times, which are related to rupture propagation and slip motion. Consequently, the separation is sensitive to the length to width ratio of the fault if the rupture and slip velocities are given their normal values. Exceptional cases of extremely slow slipping will be discussed later (cf. §5.4.4). (The intermediate trend of $\omega^{-1}$ may also be seen in the Brune model. In this case, however, they are related to the fractional stress-drop factor, $\varepsilon$, not to the propagation time; Hanks & Thatcher, 1972; see also Savage, 1972, for detailed discussion.)

### 5.4.3   Rise time

The previous discussion has introduced a new parameter, the rise time, to the study of source mechanics. A precise definition depends very much on the model to be used. For example, in the Haskell model (§5.4.1) it is defined by the transient time of the ramp function or by the time required for 90% of the final slip to occur, but in the Brune model, where an exponential function is assumed, it is defined by the time constant, $\tau$, (i.e. the time interval in which the amplitude decreases to $1/e$ of its initial value). With this difficulty in mind, let us now follow Brune (1970) and Kanamori (1974) to study this parameter in more detail.

Suppose an infinitely long fault appears instantaneously along the line AB in a medium with uniform initial stress as shown in fig. 5.15. This event may be modelled by applying negative stress to the fault surface so that the initial stress is cancelled at all points along the length of the fault. In this case, the initial motion of the medium adjacent to the fault is described with respect to a displacement $u$ and a velocity $\dot{u}$ of the following form ($y=0$ in (5.33) and (5.34)):

$$u = (\sigma/\mu)\beta t \quad \text{for } t \geq 0,$$
$$\dot{u}_0 = (\sigma/\mu)\beta \quad \text{for } t = 0.$$

As the fault is finite, however, slip will not exceed a certain limit, which

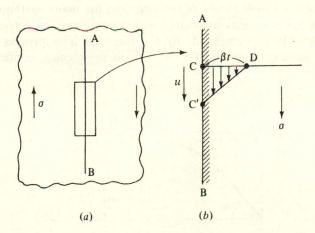

(a)              (b)

Fig. 5.15. (a) Slipping in a long fault; and (b) the particle motion on one side of the fault. (After Brune, 1970.)

is given, approximately, by $u_\infty = \sigma a/\mu$ ($a$ is nearly equal to the fault dimension) resulting in Brune's expression for $u$ and $\dot{u}_0$ (cf. (5.35) and (5.36))

$$u = u_\infty [1 - \exp(-\dot{u}_0 t/u_\infty)] \quad \text{for } t \geq 0,$$
$$\dot{u} = (\sigma/\mu)\beta \quad \text{for } t = 0,$$

where suffixes 0 and $\infty$ denote the conditions at $t=0$ and $t=\infty$, respectively. These quantities are illustrated in fig. 5.16, in which the rise time is tentatively taken as the interval $T = u_\infty/\dot{u}_0$.

The rise time may be determined directly if the ground motion in the fault vicinity is recorded completely. In the Parkfield earthquake (California, 1966, $M_S = 6.4$), for example, a strong-motion seismograph near the fault recorded the event in one component (perpendicular to the fault, see §5.5). Such a fortuitous location of a station cannot be expected for every event, but recent developments in strong-motion observation projects in several countries could considerably improve the situation.

The study of source parameters by analysis of long-period seismograms (Kanamori, 1972a) is also a promising approach to the determination of the rise time. Basically the method involves a synthesis of seismograms. The source parameters, including the rise time, are adjusted in a trial and error manner until the best fit to the observations is obtained (cf. §5.5.4). Fig. 5.17 reproduces a part of Kanamori's results for the Tottori earthquake Honshu, Japan (10 September 1943, $M = 7.4$). The figure compares several traces of the initial slope of synthetic seismograms for various rise time values. We notice that the initial slope of the record is very sensitive to this parameter. It can be seen that a rise time equal to 3 s explains the observation best.

Appendix 1 shows values of the rise time for many earthquakes. An average particle velocity associated with faulting may be derived simply by dividing the slip amplitude by the rise time. This process indicates that the velocity is about 100 cm s$^{-1}$, in most cases, confirming the

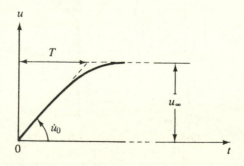

Fig. 5.16. Offset in a finite fault, as a function of time. (From Kanamori, 1974.)

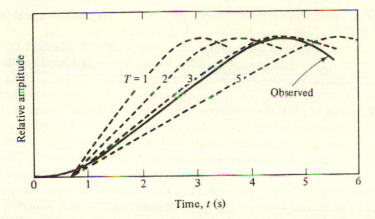

Fig. 5.17. Comparison of the initial slope of the north–south component of an observed seismogram with that of synthetic seismograms calculated for various values of rise time $T$ (given in seconds). Amplitudes are normalized. (From Kanamori, 1972a.)

previous estimate of the upper limit velocity derived from the strong-motion amplitude in the fault's vicinity in §5.4.1.

The examples given above have illustrated several important implications of the rise time for the study of physical mechanisms of earthquakes. Presumably the rise time will tell us about the initial stress level at the origin provided that (5.34) is applied appropriately to observations. If this could be done with sufficient precision, we would have a new information source for the distribution and temporal variation of crustal stresses, which are difficult to define using only conventional techniques. Rise-time analyses also play an important role in the field of earthquake engineering, in which precise prediction of future seismic vibration is sometimes requested at a critical site. As we have learnt from fig. 5.17, the initial ground velocity, and also the acceleration, are sensitively related to the rise time at the source. A preliminary application of fault-origin models to engineering problems will be discussed in §5.5.2.

### 5.4.4   *Tsunami earthquakes with extremely long process time*

Shallow submarine earthquakes of appreciable magnitudes may cause abrupt uplift or subsidence of the sea bottom over wide areas, which generates progressive water waves called *tsunamis* (or more precisely seismic tsunamis, to distinguish them from tsunamis due to other causes). Excitation and propagation of tsunamis are well described by relatively simple laws, so that we may study tsunami records at tidal stations to identify, with sufficient accuracy, patterns of off-coast vertical land movements at the epicentral area (Abe, 1973).

This special type of earthquake was discovered by Kanamori (1972*b*) who named them *tsunami earthquakes*. Examples are provided by the off-coast earthquakes in the Sanriku region, Japan (1896), and in the Aleutian region (1946). In spite of the moderate magnitudes of these earthquakes (7.5–8), the tsunamis generated were extraordinarily large, and undoubtedly in the largest tsunami group. Fig. 5.18 compares the long-period spectra in these events with those of a large but ordinary type earthquake (the Sanriku earthquake of 1933 is used as a reference).

The *effective moment*, $M_e$, in the ordinate is a hypothetical quantity which Kanamori introduced to study the spectra in a very long-period band. Let us assume a step-type displacement source function for simplicity. Then, the permanent dislocation amplitude, and consequently the moment, may be specified if a spectral density is given at any single frequency or period. If this is the case, however, conclusions obtained from different frequency bands of a given record must be consistent with each other after proper correction for the $1/\omega$ factor.

The horizontal line in fig. 5.18 for $\tau = 0$ s represents the case of a step time function. If this assumption is not made, and the time function of the Brune model with a time constant $\tau$ is assumed to apply instead then the $M_e$ values from spectral data lie on the appropriate curve in fig. 5.18. Comparison of the Aleutian and Sanriku (1896) earthquakes with the reference earthquake (Sanriku, 1933) shows that the former are characterized by $\tau = 100$ s, which is ten or more times as long as that in the reference earthquake ($\tau = 5 - 10$ s). The two shocks appear to be an order of magnitude smaller than the reference earthquake in the 20 s band which is most critical for magnitude determination. In a much longer period band, 500 s for example, all three earthquakes appear to be about the same size, suggesting equal potential for tsunami generation.

Fig. 5.19 illustrates the basic properties of tsunami earthquakes. Case B refers to the above examples ($\tau = 10^2 - 10^3$ s), whereas case A represents ordinary type earthquakes ($\tau = 1 - 10$ s). Partitions of seismic energy in short- and moderate-period waves are compared in the right half of the picture. The physical mechanisms that produce such extremely slow focal processes are unknown at present. But it is very likely that some anelastic, presumably viscous or plastic, properties in the Earth's medium play important roles in these processes. Extrapolating to an even longer rise time, the hypothetical case C, suggesting a 'silent' earthquake, may be drawn. Fault creep and migration of anomalous crustal strains along a fault (§7.3) might be considered as extreme cases. (Recently, Fukao (1979) pointed out that large tsunamis can be accounted for by greater ocean-bottom vertical movements within accretionary prisms on steeply dipping thrust faults, rather than by an anomalously long process

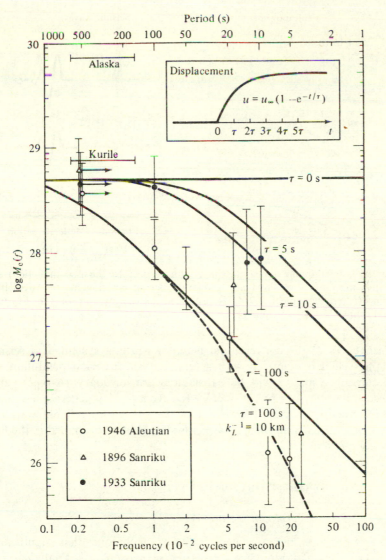

Fig. 5.18. Effective moment as a function of frequency. The insert shows a dislocation function with a time constant introduced to fit the data. The solid curves give the effective moment corresponding to the dislocation functions with $\tau = 0$, 5, 10 and 100 s. The dashed curve is for a finite propagating source having a time constant of 100 s and a correlation length of 10 km. (From Kanamori, 1972*b*.)

time as Kanamori originally supposed. These two possibilities could be examined if broad-band records of seismic waves and precise patterns of sea-bottom movements are available in future earthquakes.)

Discovery of tsunami earthquakes has raised serious doubts about the

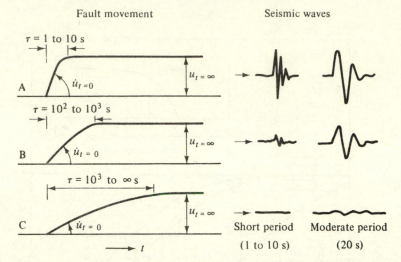

Fig. 5.19. Earthquake types classified with respect to the rise time (left) and relative amplitudes of short and moderate period components of waves from them. (After Kanamori, 1974. Reproduced by permission of the Maruzen Publishing Co., Tokyo, Japan.)

effectiveness of the conventional tsunami warning system. If a seismograph system is equipped with a relatively-short-period pendulum, the instrument response will be satisfactory for ordinary (A-type) earthquakes, but may not be satisfactory for events which generate few short-period waves, such as those of type B. For more reliable warnings, therefore, the existing warning systems should be re-examined in the light of this possibility.

### 5.5  SYNTHESIS OF SEISMIC EFFECTS

#### 5.5.1  *General considerations*

Recent progress in computer-simulation techniques has significantly changed the methodological structure of seismology. Analytical procedures have been a principal method for research in seismology since its early days. That is to say, observational data were analysed and only primary information, in a form simple enough for comparison with theories, was extracted from them. The preparation of a travel-time curve may be taken as an example. The times of onset of waves were recorded, disregarding all the other information on the seismogram, and were compiled into a travel-time chart giving a summary of wave propagation. This was the only practical approach in the days before computers were available.

Recent developments in theory and in computer techniques have made an alternative approach possible. Using computers, we can construct a fairly realistic seismic model and let it work numerically under plausible conditions. In other words, we can construct a model and compare the responses of the model directly with observation.

The introduction of computer-simulation techniques into seismology is advantageous not only for basic research, but also for the application of theory to practical problems. A precise synthesis of seismic effects will warn us of possible vibrations at a given site and greatly help the design of critical structures (such as large buildings, long bridges, oil refineries, power stations, water reservoirs, etc.) to be built on that site. Accuracy and precision of syntheses, which depend on our knowledge of the physical processes involved and on the accuracy of seismic parameters, is not yet sufficient for this purpose. But, this technique will be continuously improved if seismology continues to advance at its present rate.

### 5.5.2 *Strong-motion spectra of earthquakes*

The ground vibration in the vicinity of a fault-origin has only been reached in a few fortunate cases when a seismic instrument with the proper dynamic range has been operating close to the activated portion of a fault. A principal tool for this purpose is a strong-motion seismograph, originally developed for earthquake engineering. A strong-motion seismograph, though several different models are available, is basically an accelerometer, which gives direct readings of the acceleration of the ground motion in the ordinary period range of seismic waves.

In the discussions so far, we have occasionally used the terms near or far field, without any specific definition. These are used to specify the magnitude of the epicentral distance relative to the source dimension and wavelength to be studied. The term *far field* applies to epicentral distances which are large in comparison with these parameters, whereas the term *near field* applies to distances much shorter than them. Some of the following examples will discuss S-waves with dominant periods of 0.1–10 s (0.3–30 km in wavelength) at distances of 10–50 km from faults of moderate sizes. An additional term, an *intermediate field* may sometimes be necessary to specify these cases.

Hasegawa (1974), using the theory proposed by Savage (1972), synthesized strong-motion spectra for several major earthquakes and compared them with observed data using records at distances of 10–50 km. The theory by Savage was originally intended for the study of far-field problems (§5.4.2), but was applied by Hasegawa in preference to the near-field theory developed by Haskell (1969) which could also have been used. Fig. 5.20 shows a schematic view of Hasegawa's model, using symbols previously defined in §5.4.2. Applying the expressions given in

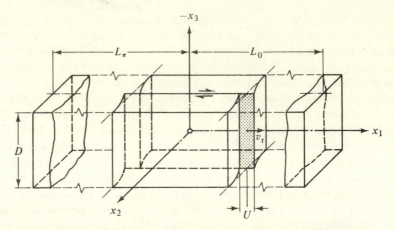

Fig. 5.20. Fault parameters associated with a fault movement located about the coordinate axes shown in fig. 5.13. (From Hasegawa, 1974.)

§5.4.2, we can write the spectral amplitude of the S-wave, which predominates in seismograms under the present conditions, as:

$$|\hat{\mathbf{u}}_S| = \mathbf{R}_S(\theta, \phi, r) M_0 \omega |\hat{G}| F_2(\omega, \tau_0, \tau_\pi),$$

with $|\hat{G}|$ defined by

$$|\hat{G}| = (1/\omega)/(1 + \omega^2 \tau^2)^{1/2},$$

and $M_0 = \mu U L D$ as before. This spectrum has three linear trends and two corner frequencies, $\omega_1$ and $\omega_2$, as shown in fig. 5.14.

The spectral density of ground acceleration can be calculated from the spectral density of displacement after correction for the $1/\omega^2$ factor. Taking into consideration attenuation and the free surface effect, the acceleration amplitude spectral density, $S$, is given by

$$S = 2|\hat{\mathbf{u}}_S|\omega^2 e^{-kr}, \quad (\text{cm s}^{-1}) \tag{5.50}$$

where $k$ is the attenuation coefficient, $r$ is the travel path length (hypocentral distance) and the factor 2 represents the amplifying effect at the free surface (note that the unit of acceleration spectral density is cm s$^{-1}$ instead of cm s which is the unit of the displacement spectral density). The attenuation coefficient $k = \omega/2Q\beta$, where $1/Q$ is the specific dissipation factor and $\beta$ is the shear wave velocity. Suppose the duration time of the transverse horizontal component of the acceleration seismogram, $a(t)$, is equal to $T$, then the Fourier transform of the observed ground acceleration is

$$S = \left| \int_0^T a(t) e^{-i\omega t} dt \right| \quad (\text{cm s}^{-1}) \tag{5.51}$$

which corresponds to the strong-motion spectrum in earthquake engineering (strong-motion seismographs are mostly designed to record ground acceleration in the ordinary seismic band; see §1.3 and fig. 1.4).

Hasegawa (1974) synthesized spectra for several earthquakes for which strong-motion records or spectra are available. Fig. 5.21 illustrates this technique, step by step, for the case of an earthquake of magnitude 5.5, located at a distance of 10 km from the station. Table 5.1 gives average values of the parameters $M_0$, $L$, $D$, and $U$ as functions of $M$. Suppose the fault is a bilateral fracture initiated at a point dividing the fault length in the ratio $1:2$, then the basic parameters are as shown at the top of fig. 5.21 where $v_r$ is assumed to be 3.4 km s$^{-1}$. The top curve in fig. 5.21 represents $|\hat{u}_s|$ calculated by the equation given above. After correction for the frequency factor and for the attenuation during propagation, we finally obtain the bottom trace for (5.50), where the $Q$ factor is varied producing a family of curves.

Comparisons of synthetic with observed spectra may be seen in fig. 5.22. Fig. 5.22a shows an excellent fit for the Imperial Valley (or El Centro) earthquake of 19 May 1940 ($M = 7.1$), justifying the theory and the parameters used in the present analysis. An example of a poor fit is shown in fig. 5.22b for the Borrego Mountain, California, earthquake (9 April 1968, $M = 6.5$–6.7). Disagreement is significant in the lower frequency range, with the theoretical spectrum about one order-of-magnitude less than the experimental one. Hasegawa suggested, after preliminary tests, that the fit could be improved if the contribution from Love waves is included (see Hasegawa, 1974, for further discussion).

Leaving these problems for future consideration, however, experience so far seems to prove the usefulness of the synthesis technique. As this method is tested and improved through the analysis of more earthquakes, it should become possible to predict strong-motion spectra in future earthquakes with enough accuracy for useful practical application.

### 5.5.3  Seismic displacement near a fault

*Simulation of fault movement.* Recent studies of far-field problems have enabled us to determine source parameters from remote station recordings. More direct information on fault movement can be obtained if reliable observations are made in the vicinity of a fault origin, but this is not always possible. One of the rare and fortunate cases of reliable near-fault observation occurred for the Parkfield, California, earthquake of 1966, when a strong-motion seismograph recorded the ground motion at a distance of only 80 m from the activated portion of the San Andreas fault. Aki (1968) used data from this seismograph to prove that the observed ground motion is in good agreement with the theoretical

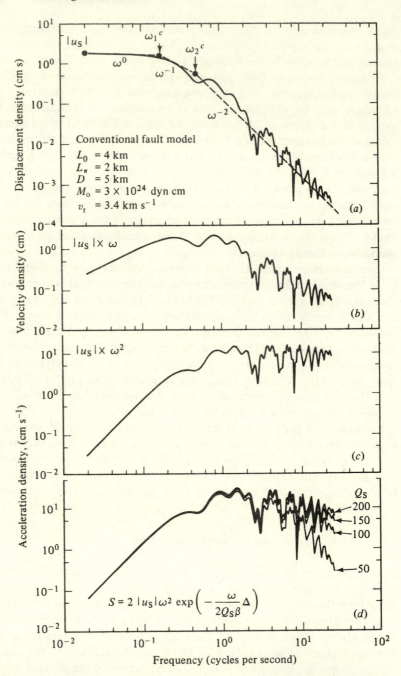

Fig. 5.21. Step-by-step generation of the theoretical Fourier amplitude spectrum, $S$, for an earthquake of magnitude 5.5 located at a distance of 10 km from the station. The effect of varying $Q_S$ is illustrated in part $(d)$. (From Hasegawa, 1974.)

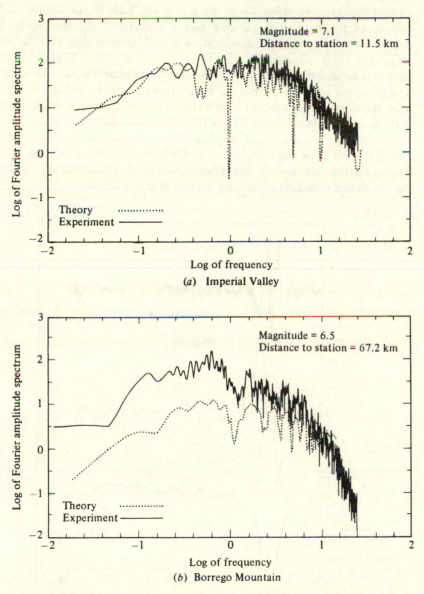

Fig. 5.22. Examples of: (*a*) a good fit between theoretical and observed Fourier amplitude spectra at El Centro, California – for the Imperial Valley earthquake of 19 May 1940; and (*b*) a poor fit for the Borrego Mountain earthquake of 9 April 1968. Frequencies are measured in cycles per second and spectra in cm s$^{-1}$. (After Hasegawa, 1974.)

ground motion which is simulated by taking a set of likely fault parameters.

A part of the strong-motion record, which in practice represents the ground acceleration, is shown in the top of fig. 5.23. In the middle and bottom of the figure the velocity and displacement are shown, both calculated from the original acceleration record by numerical integration. Note that these components are in the direction perpendicular to the fault extension; unfortunately, the parallel component was not recorded. The principal part of the motion is surprisingly simple, as can be seen from the displacement record which is well represented by a single pulse with an amplitude of about 30 cm (N65°E) and a duration of 2–3 s.

To calculate the near-field displacement, Aki referred to Maruyama (1963) for the basic expressions for a displacement field produced by a dynamic dislocation source. Maruyama gives the displacement $u(Q, t)$ at $Q$ due to dislocation $U(t)$ on a fault surface element $d\Sigma$ as

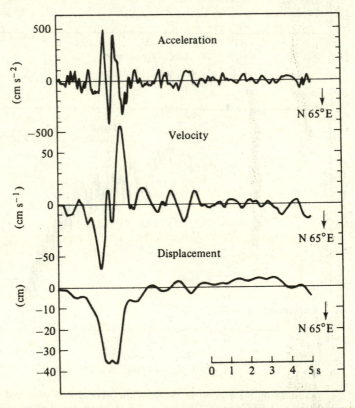

Fig. 5.23. Ground velocity and displacement obtained by integrating the perpendicular component accelerogram for the Parkfield–Cholame earthquake of 1966. (From Aki, 1968.)

$$u_m(Q,\,t)=\frac{\mathrm{d}\Sigma}{4\pi\rho}\sum_{l=1}^{3}\nu_l$$

$$\times\sum_{k=1}^{3}\left\{6\mu\left[-\delta_{kl}\frac{r_m}{r^5}-\delta_{mk}\frac{r_l}{r^5}-\delta_{lm}\frac{r_k}{r^5}+5\frac{r_k r_l r_m}{r^7}\right]\left[\Psi_k\!\left(t-\frac{r}{\alpha}\right)-\Psi_k\!\left(t-\frac{r}{\beta}\right)\right]\right.$$

$$+6\mu\left[-\delta_{kl}\frac{r_m}{r^4}-\delta_{mk}\frac{r_l}{r^4}-\delta_{lm}\frac{r_k}{r^4}+5\frac{r_k r_l r_m}{r^6}\right]\left[\frac{1}{\alpha}\Psi_k'\!\left(t-\frac{r}{\alpha}\right)-\frac{1}{\beta}\Psi_k'\!\left(t-\frac{r}{\beta}\right)\right]$$

$$+\left[(\lambda-2\mu)\,\delta_{kl}\frac{r_m}{r^3}-2\mu\,\delta_{mk}\frac{r_l}{r^3}-2\mu\,\delta_{lm}\frac{r_k}{r^3}+12\mu\frac{r_k r_l r_m}{r^5}\right]\left[\frac{1}{\alpha^2}\Psi_k''\!\left(t-\frac{r}{\alpha}\right)\right]$$

$$+\left[2\mu\,\delta_{kl}\frac{r_m}{r^3}+3\mu\,\delta_{mk}\frac{r_l}{r^3}+3\mu\,\delta_{lm}\frac{r_k}{r^3}-12\mu\frac{r_k r_l r_m}{r^5}\right]\left[\frac{1}{\beta^2}\Psi_k''\!\left(t-\frac{r}{\beta}\right)\right]$$

$$+\left[\lambda\,\delta_{kl}\frac{r_m}{r^2}+2\mu\frac{r_k r_l r_m}{r^4}\right]\left[\frac{1}{\alpha^3}\Psi_k'''\!\left(t-\frac{r}{\alpha}\right)\right]$$

$$+\left.\left[\mu\,\delta_{mk}\frac{r_l}{r^2}+\mu\,\delta_{lm}\frac{r_k}{r^2}-2\mu\frac{r_k r_l r_m}{r^4}\right]\left[\frac{1}{\beta^3}\Psi_k'''\!\left(t-\frac{r}{\beta}\right)\right]\right\},\tag{5.52}$$

where $u_m$ is the $m$th component of $u$, $k=(1,2,3)$, $l=(1,2,3)$, $m=(1,2,3)$, $\nu_i$ is the $i$th component of the normal to $\mathrm{d}\Sigma$, $\Psi_k(t)$ is the component of the double-integral of the dislocation:

$$\Psi(t)=\int_0^t \mathrm{d}t'\int_0^{t'}\Delta U(t'')\mathrm{d}t'',\tag{5.53}$$

$\alpha$ and $\beta$ are the compressional and shear wave velocities respectively, and $\lambda$, $\mu$ are Lamé's constants ($\delta_{kl}=0$ if $k\neq l$; $\delta_{kl}=1$ if $k=l$). In the case of the Parkfield earthquake, the normal to $\mathrm{d}\Sigma$ is perpendicular to $U$ and $\delta_{kl}$ vanishes, so that the above expression may be shortened to some extent. $\Psi'$, $\Psi''$ and $\Psi'''$ represent the first, second and third derivatives of $\Psi$ with respect to time, respectively.

Integration of $u(Q,t)$ over the whole fault surface may be done numerically; that is, we divide the whole area into many small elements of uniform area $\Delta d\times\Delta l$, and sum all the individual contributions $u(Q,t)$; see fig. 5.24.

In order to carry out the numerical calculation, we tentatively assume a step time function $\Delta U(t)$, with its Fourier transform given as

$$\Delta U(\omega)=U/i\omega,\tag{5.54}$$

where $U$ is the final dislocation amplitude. Then (5.52) may be written as

$$u(Q,\,\omega)=\int_{-\infty}^{\infty}u(Q,\,t)e^{-i\omega t}\mathrm{d}t=[Ae^{-i\omega r/\alpha}+Be^{-i\omega r/\beta}]U\mathrm{d}\Sigma,\tag{5.55}$$

where $A$ and $B$ are complicated functions of $\omega$, $r$, $\alpha$, and $\beta$.

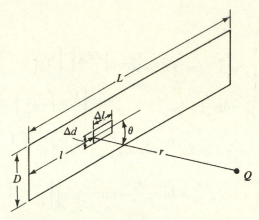

Fig. 5.24. Geometry of Aki's fault model. (From Aki, 1968.)

The effects of dislocation propagation along the fault length, $L$, with velocity $v_r$ may be represented by the Ben-Menahem transfer function for each fault section as:

$$Z_c = [(\sin X_c)/X_c]e^{-iX_c} \tag{5.56}$$

where $X_c = (\omega\Delta l/2c)[(c/v_r) - \cos\theta]$, and $c$ stands for the velocity $\alpha$ (P-wave) or $\beta$ (S-wave) as in §5.1.2. Assigning each contribution with a phase delay due to rupture propagation with velocity $v_r$ for distance $l$ from the first break point (assumed at the northern end of the fault trace), we obtain:

$$u(Q, \omega) = U\Delta d\Delta l \left[ A\frac{\sin X_\alpha}{X_\alpha} \exp\left(-i\omega r/\alpha - iX_\alpha - i\omega l/v_r\right) \right.$$
$$\left. + B\frac{\sin X_\beta}{X_\beta} \exp\left(-i\omega r/\beta - iX_\beta - i\omega l/v_r\right) \right] \tag{5.57}$$

The displacement at the station, $Q$, is obtained by summing the displacements expressed by (5.57) from all elementary sections.

Further corrections must be applied to the integrated result to account for several factors, such as the variation of $U$ in the vertical direction in the fault (see for example (4.6) in Knopoff's model) and wave amplification at the free surface.

Geographically, the Parkfield earthquake fault appeared as a series of fracture zones, extending from northwest to southeast, all striking in the same direction with uniform right-lateral slips. In order to simplify the calculation, Aki (1968) represented the fault as a system of five fault segments, NW3, NW2, NW1, SE1 and SE2, from northwest to southeast. In practice, the last four segments belong to a single fracture zone, which was divided into the NW and SE groups at a point near the middle closest to the seismic station. The segments NW1 and SE1 each represent a 1 km extension closest to the station. Lengths of the other segments, NW3, NW2 and SE2 are 30, 4 and 1 km, respectively, so that the total length of the five amounts to 37 km.

Fig. 5.25a shows the synthesized displacement parallel to the fault, although we have no observational data for comparison. However, we notice that the trace shows a step-wise displacement toward the northwest at time $t_0$ when the rupture passes the nearest point to the station. This is reasonable in view of the right-lateral sense of slip on the fault and shows the usefulness of the synthesis technique. An amplitude of $0.66 \times U/2$ seems small in comparison with the theoretical amplitude at the fault's edge, $U/2$. Separation of the station from the fault and the cut-off of higher frequency components at 3 cycles per second during calculation are suspected causes of this reduced amplitude.

There is a clear agreement between the synthetic pulse form with observation in the perpendicular component (compare figs. 5.25b and 5.23). To explain the observed amplitude (approximately 30 cm), however, $U$ must be approximately 60 cm. This is more than ten times as large as the fault offset at the surface immediately after the earthquake near the station site (Allen & Smith, 1966). Aki examined

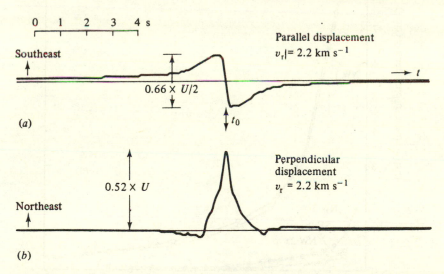

Fig. 5.25. The synthetic seismograms obtained by summing the contributions from all significant fault segments. The value $U$ indicates the surface value of a given dislocation. The parallel component shows a step-like motion directed toward the northeast, as expected from the given boundary condition. The perpendicular component shows a pulse-like motion directed toward the northeast, in agreement with the observations. (From Aki, 1968.)

several possibilities to account for this disagreement in the offset amplitude.

One possible explanation is the presence of a soft superficial layer which conveys the base rock offset to the surface with its amplitude reduced.

Teleseismic data gave a seismic moment of $1.4 \times 10^{25}$ dyn cm for the Parkfield earthquake. Suppose we take the parameters associated with this event as $L = 30\text{–}37$ km, $\mu = 3 \times 10^{11}$ dyn cm$^{-2}$ and $U = 50$ cm respectively, then the fault depth, $D$, in (5.23) is only 3 km, which is extremely small when compared with the value of 15 km indicated by the aftershock distribution (Eaton, 1967). To explain this disagreement, one may attribute a shallow slip to the main shock and a deeper slip to the aftershocks, though there is as yet no evidence to support this explanation (Aki, 1968).

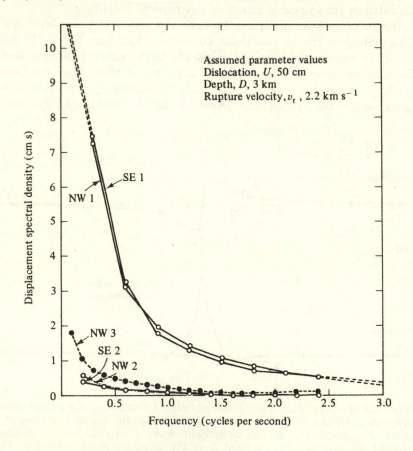

Fig. 5.26. The displacement spectral density for the parallel component obtained by Aki's method. The contributions from several fault segments are shown separately. (From Aki, 1968.)

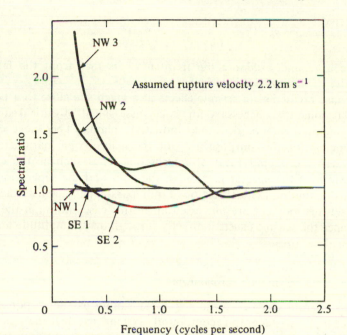

Fig. 5.27. The ratio of seismic spectral density corresponding to a fault with 13 km depth to one with 3 km depth. The ratio must approach $13/3 = 4.33$ at long periods and at large distances. The ratio is, however, close to unity because of the short distances of the different fault segments to the station. (From Aki, 1968.)

*The dominant contribution of the fault segment closest to near-field ground motion.* Comparison of the individual contributions from fault segments to the near field has shown that seismic effects in the fault's vicinity are mainly responses to the motion of the nearest portion of the fault. Fig. 5.26 compares the displacement spectral density (parallel component) contributed by the individual fault segments described in the previous section. Notice that the contributions from NW1 and SE1 predominate in the present case, in spite of their short lengths. The other three segments give small contributions although they are predominant components in the teleseismic field. This fact suggests that seismic effects in the fault vicinity are due not to the whole fault surface, but to a limited area in the fault which is adjacent to the observation site. A similar relation is noticed when considering the vertical extension of a fault. Fig. 5.27 illustrates ratios of spectra of two models, with depths of 13 km (as in the Parkfield earthquake) and 3 km. The ratio must approach $13/3 = 4.33$ at long periods and at long distances. However, it remains at unity for a considerable range, especially in the cases of NW1 and SE1, suggesting that the ground motion at the present site is due mainly to faulting in the upper few kilometres. Aki introduced the following inequality, which allows us to estimate a critical depth $D_c$ which is the depth below which the contribution from faulting can be neglected:

$$(\omega/cr)^{1/2}(2D_s/\pi) \gg 1. \qquad (5.58)$$

where $\omega$, $r$ and $c$ denote the frequency, distance from the fault and velocity of the wave, respectively.

Precise prediction of seismic effects at a given site close to a potential fault is sometimes necessary for safety assessment of critical structures like nuclear reactors or other industrial plants. The most essential parameters for this purpose are not the macroscopic parameters of a supposed source, but local or microscopic ones such as the expected behaviour of the nearest portion of the fault and the separation of the site from this particular portion. Fault parameters such as magnitude and seismic moment are not essential in this case, although they might influence the seismic effects indirectly through offset amplitude and other physical parameters.

### 5.5.4   *Synthetic seismograms*

A seismogram can be considered most simply as the response of the source–path–seismograph compound system to disturbances at the source. If the transfer functions are known, therefore, we may synthesize a seismogram from a signal assumed at the source. The first work of this kind (known as 'Lamb's problem', see Love, 1944) modelled the fundamental structure of a seismogram using a simple model of an earthquake.

Progress in source-mechanics studies allows us to deal with more precise models. For example, Kanamori (1972a) used the theory of Haskell (1969) to synthesize seismograms for the Tottori (1943) earthquake and compared them with the observed seismograms recorded at the Abuyama Seismological Observatory, located about 140 km southeast of the epicentre (fig. 5.28). As this is a vertical strike-slip fault, its displacement field on the surface of the semi-infinite Earth may be calculated from Haskell's model in an infinite medium, where the width of the surface displacement should be taken as twice that of the value in the infinite medium. In his calculations, Kanamori used the values of fault length, $L$, fault width, $D$, average slip, $U$, and the ramp-type source function with rise time $T$ given in the figure (cf. §5.4.3. for $T$), and compared the synthesized seismograms with observation (fig. 5.28 illustrates the traces after correction for the instrument response). Disregarding relatively-short-period waves on the record, there is excellent agreement of observation with the synthetics, especially for the N–S component of waves due to bilateral faulting with $v_r = 2.3$ km s$^{-1}$. The analysis of the E–W component supports this conclusion. The final set of fault parameters, which provides the best fit with geodetic and seismological data is:

| | |
|---|---|
| length, $L$ | 33 km, |
| depth, $D$ | 13 km, |
| strike | N80°E, |
| rupture-propagation velocity, $v_r$ | 2.3 km s$^{-1}$ (for the bilateral type), |
| average dislocation, $U$ | 2.5 m, |
| seismic moment, $M_0$ | $3.6 \times 10^{26}$ dyn cm, |
| rise time, $T$ | 3 s, |
| particle velocity, $u_0$ | 42 cm s$^{-1}$ (on the fault surface). |
| stress-drop, $\Delta\sigma$ | 83 bar, |
| effective tectonic stress, $\sigma$ | 30–100 bar. |

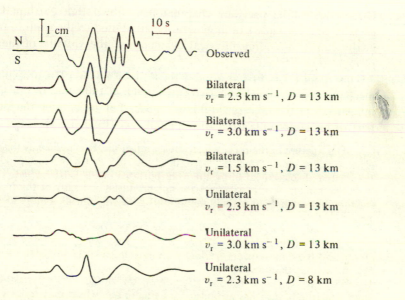

Fig. 5.28. Comparison of the observed N–S component seismogram with synthetic seismograms calculated from various fault models. The rise time, $T$, the dislocation, $U$, and the fault length, $L$, are held constant with values of 3 s, 2.5 m and 33 km respectively. The mode of rupture propagation, the rupture velocity and the depth of fault, $D$, are varied (From Kanamori, 1972$a$.)

# 6 Physics of focal processes

### 6.1.1 *The elastic rebound theory*

Discussion in the previous chapters has shown that earthquakes, especially shallow ones, can in general be attributed to the rupture of faults in the Earth, providing physical evidence for the validity of the *elastic rebound theory*.

This theory was first proposed by H. F. Reid in 1910 to explain the mechanism of the San Francisco earthquake of 1906, and has profoundly influenced earthquake seismologists since. Let us extract the principal part of his statements from the original text (Reid, 1911):

1. The fracture of the rock, which causes a tectonic earthquake, is the result of elastic strains, greater than the strength of the rock can withstand, produced by the relative displacement of neighboring portions of the earth's crust.
2. These relative displacements are not produced suddenly at the time of the fracture, but attain their maximum amounts gradually during a more or less long period of time.
3. The only mass movements that occur at the time of the earthquake are the sudden elastic rebounds of the sides of the fracture towards positions of no elastic strain; and these movements gradually diminishing, extend to distances of only a few miles from the fracture.
4. The earthquake vibrations originate in the surface of fracture; the surface from which they start has at first a very small area, which may quickly become very large, but at a rate not greater than the velocity of compressional elastic waves in the rock.
5. The energy liberated at the time of an earthquake was, immediately before the rupture, in the form of energy of elastic strain of the rock.

These statements are called the *elastic rebound theory of tectonic earthquakes*; they do not broach furthermore the original cause of earthquakes, which lies in the source of the slow movements accumulating the elastic energy, but merely give the *modus operandi* of the accumulation and liberation of this energy.

### 6.1.2 *Static and dynamic aspects of a seismic origin (summary)*

Let us review the previous five chapters and enumerate the most essential observational evidence supporting the validity of the fault-origin hypothesis:

(*a*) The seismometrically located hypocentre of an earthquake falls

126

consistently on the surface of the fault-origin model derived from land movement data (§4.1.3).

(*b*) The observed static and dynamic displacement fields are success-fully explained by a double-couple force system at the source (§§3.1, 3.2, 3.3 and §§4.2 and 4.3; see also the theorem of force–dislocation equiva-lence in §4.3.2).

(*c*) Seismic energy release and strain-energy change are observed to be of the same order of magnitude (see data in appendix 1).

(*d*) Source geometry and mechanical parameters are consistent with the concept of an earthquake source volume (§2.3, see also scaling laws in §6.1.3).

(*e*) Observed far-field displacement spectra are consistent with a step time function for the fault slip (see several modified source time functions in §§5.2 and 5.4), and good agreement is obtained between the static and dynamic moments at a source (§5.3).

(*f*) Observed phase shifts and macroseismic aspects in the epicentral area are well explained by rupture-propagation effects (§5.5). Location of an epicentre, which tends to appear near an extremity of a source area, may be well understood if the epicentre (more precisely, the hypocentre) represents a point of first breaking.

Note that these features may be comprehensively explained by a unifying theory, if a seismic source is considered to be a finite surface of shear fracture propagation. No other view of earthquake sources ac-counts for them better. Therefore, we may confidently adopt the idea that earthquakes are generally attributed to the rupture of faults in the Earth, as earlier speculated by Reid.

We must remark here that our present understanding of the nature of earthquakes is mainly founded on information from large shallow earth-quakes. In other words, it does not apply immediately to other types of events, such as deep earthquakes and very small earthquakes. Also, the previous chapters did not discuss the physical mechanisms and the tectonic causes of earthquakes in detail. These topics will be discussed in more detail in this and the following chapters.

A schematic view of an earthquake source is shown in fig. 6.1*b* together with a sketch of the ground surface movement along the fault's extension (fig. 6.1*a*). Notice that a considerable number of parameters are employed, even in a relatively simple model like this. They are:

| | |
|---|---|
| *fault geometry* | length $L$; depth (bottom) $D$, (top) $d$; strike $\theta$; dip $\delta$; |
| *mode of slip* | slip vector $U(t)$; rise time $T$; |
| *rupture propagation* | location of first breakage $x_1^*$, $x_2^*$, $x_3^*$; rupture velocity $v_r$. |

In addition to these primary parameters, several compound parameters, stress-drop $\Delta\delta$, moment $M_o$, energy $E$, are also used as discussed in the

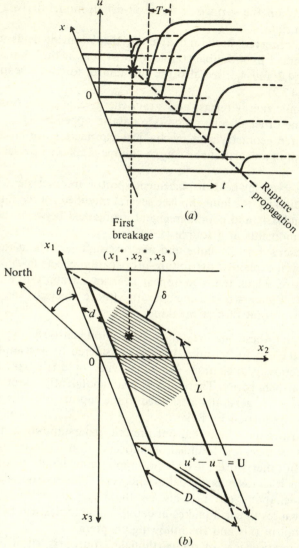

(a)

(b)

Fig. 6.1. (*a*) Displacement at the ground surface above a fault; and (*b*) fault geometry for a fault origin with rupture propagation (shaded).

previous two chapters. Examples of source parameters are compiled in appendix 1.

### 6.1.3   *Scaling laws*

A number of source parameters have been introduced in the previous discussion. Some of them are considered, in principle, to be independent observable quantities (e.g. fault geometry and slip amplitude), though others are compound (e.g. moment and energy) and depend on several basic parameters in their definitions. In this sense, the mechanical system of the seismic source may appear to have a considerable number of degrees of freedom.

In an actual seismic event, however, the number of degrees of freedom is small, since the parameters, even those formerly thought to be independent, are empirically interrelated. In other words, they can take values only in a restricted manner. This situation is illustrated most clearly if we accept a similarity between large and small earthquakes, following Aki (1967, 1972).

He assumed that earthquakes are geometrically similar, and that stress-drop, rupture and slip velocities are independent of magnitude, from which it follows that the corner frequency is inversely proportional to the fault length (see for example (5.49)), and the seismic moment proportional to the cube of fault length. Using the Haskell model (§5.4.1), he represented the statistical form of radiated seismic spectra by the $\omega$-*square model* (or more precisely the $\omega^{-2}$ model), which is flat for frequencies lower than the corner frequency, and falls off with a gradient of $\omega^{-2}$ for higher frequencies (see fig. 5.14).

With these assumptions, Aki (1967) synthesized the far-field spectra shown in fig. 6.2. The family of curves in the figure are arranged in such a manner as to satisfy the conditions: (*a*) the corner frequency is inversely proportional to the fault length, i.e. to the cube root of the seismic moment (i.e. the broken line in fig. 6.2 has a trend given by $T^3$); and (*b*) the spacing between the curves is uniform at the period of 20 s (vertical arrow in fig. 6.2) to be compatible with the definition of $M_S$ given by Gutenberg & Richter (§2.1).

If this scaling law holds for a wide range of earthquake magnitudes, it implies fundamental relations between source parameters. In fact, Aki tested several empirical formulae relating source parameters to magnitude and showed that his model predicts these relations very well. In detail, however, he noticed considerable disagreement of the law from observation, especially for periods shorter than 10 s. His later models (Aki, 1972) modified the $\omega$-square model for this frequency range.

An inclusive interpretation of the empirical dependence of source parameters on earthquake scales has been attempted by Kanamori & Anderson (1975), on the basis of simple crack and dynamic dislocation models. Let us first assume the following similarities:

$$\left.\begin{array}{l} D/L = c_1 = \text{constant} \quad \text{(geometrical similarity)}, \\[4pt] U/L = c_2 = \text{constant} \quad \text{(strain-drop, fixed)}, \\[4pt] v_r T/L = c_3 = \text{constant} \quad \text{(dynamic similarity)}. \end{array}\right\} \quad (6.1)$$

The first two equations imply that $M_o = \mu U A \approx L^3$. The magnitude $M_S$ is

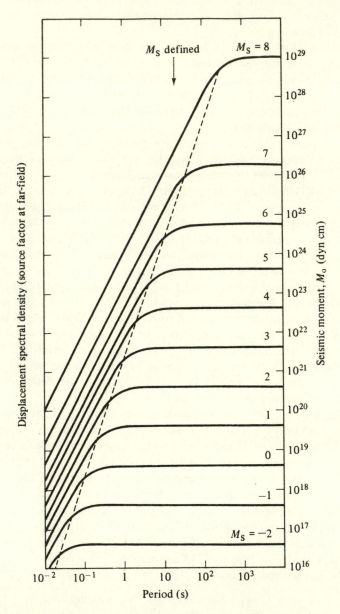

Fig. 6.2. Far-field seismic spectral density from earthquakes with various magnitudes, $M_s$, and moments, calculated using the $w$-square model. (After Aki, 1967, 1972.)

proportional to the logarithm of the far-field displacement spectral amplitude at $\omega = \omega_0 = 2\pi/T_0$ ($T_0 = 20$ s, by definition), and is given approximately by

$$M_S \approx \log \left[ ULD \left| \frac{\sin \frac{\omega_0 T}{2}}{\frac{\omega_0 T}{2}} \right| \left| \frac{\sin \frac{\omega_0}{2} \frac{L}{v_r}}{\frac{\omega_0}{2} \frac{L}{v_r}} \right| \right] \approx \log L^3 + \log [F(T, L/v_r)],$$

(6.2)

on the basis of the Haskell model (see §5.4.2). The function $F(T, L/v_r)$ takes the following asymptotic values depending on $T$ and $L/v_r$:

$$F(T, L/v_r) \approx \begin{cases} (v_r T_0)/(\pi L) \approx L^{-1} & \text{for } T < T_0/\pi \text{ and } L/v_r > T_0/\pi, \\ (T_0/\pi T) \approx L^{-1} & \text{for } T > T_0/\pi \text{ and } L/v_r < T_0/\pi, \\ (T_0^2 v_r)/(\pi^2 LT) \approx L^{-2} & \text{for } T > T_0/\pi \text{ and } L/v_r > T_0/\pi, \\ 1 \approx L^0 & \text{for } T < T_0/\pi \text{ and } L/v_r < T_0/\pi. \end{cases}$$

Therefore

$$M_S \approx \begin{cases} \log L^2 & \text{for } T < T_0/\pi \text{ and } L/v_r > T_0/\pi, \\ \log L^2 & \text{for } T > T_0/\pi \text{ and } L/v_r < T_0/\pi, \\ \log L & \text{for } T > T_0/\pi \text{ and } L/v_r > T_0/\pi, \\ \log L^3 & \text{for } T < T_0/\pi \text{ and } L/v_r < T_0/\pi, \end{cases}$$

(6.3)

Fig. 6.3 shows the dependence of $M_S$ on $L$ and of $M_o$ on $M_S$ in the four quadrants of the $(T, L/v_r)$ plane and also shows those earthquakes for which the rupture velocity is known. Notice that most earthquakes ($M_S$: 6.5–7.5) in this example fall on the quadrant for $M_S \approx \log L^2$. This gives:

$$\log M_o \approx \tfrac{3}{2} M_S.$$

(6.4)

As discussed in §5.4.2, the seismic source process is associated with two kinds of characteristic time, i.e. a rupture-propagation time ($L/v_r$) and a rise time ($T$). We can, in fig. 6.3, compare these process times. Vertical and horizontal thick reference lines are shown in the figure for a reference time duration of $20/\pi (\approx 6)$ s. By this standard, most earthquakes have 'rapid' rise-times, for which $\log M_o \approx 1.5\, M_S$ or $\log M_o \approx M_S$. If $T$ is extraordinarily long, as is the case for tsunami earthquakes (§5.4.4), then $\log M_o \approx 3 M_S$. Several seismologists have noticed that the $\log M_o$ versus $M_S$ curve tends to slope more steeply for very large magnitude events (see for example Aki, 1972; Ohnaka, 1976). This is also understandable from the present viewpoint, since the rise time may be significantly longer in very large earthquakes.

Magnitude is the most popular measure of earthquake size. It is interesting that a single number, derived from such a simple procedure,

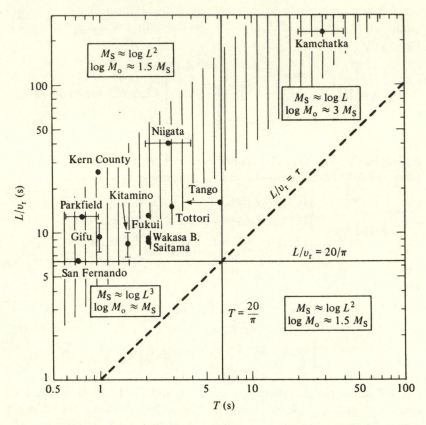

Fig. 6.3. Relations between rupture time, $L/v_r$, and the rise time $T$. The dependence of the magnitude $M_S$ on the linear source dimension, $L$, and the seismic moment $M_0$ is shown for each quadrant. Observed earthquakes are distributed along the shadowed zone. (After Kanamori & Anderson, 1975.)

can represent the overall physical size of a wide range of earthquakes, from very small earthquakes to disastrous earthquakes. For extremely large earthquakes, however, there are considerable uncertainties in the results, even when the surface wave magnitude, $M_S$ is used. Let us compare two big earthquakes, the Sanriku (1933) and Alaska (1964) earthquakes of similar magnitude ($M_S = 8.3$–$8.5$). In terms of the fault parameters, the latter is much bigger, about 9 times as large in terms of the fault surface area, and about 6 times as large in terms of the offset, and consequently about 56 times as large in terms of the seismic moment $M_0$ (see appendix 1). In other words, the $M_S$ scale tends to 'saturate' and lose its validity for earthquakes beyond magnitude 8.

This saturation of the $M_S$ scale may be attributed to extremely large source dimensions, generating extremely long-period seismic waves in these events. Hence the previous 20 s surface waves are only incidental. The seismic moment is the most useful parameter in such cases as it is an excellent scaling factor in all equations for the dynamic radiation of seismic waves. Also, it is a scale of clearcut physical definition, related in a simple manner to the static fault parameters (recall that the magnitude scale is essentially a relative scale). For these reasons, the seismic moment has recently been widely used in seismology, especially in long-period seismology.

If a proper conversion from the seismic moment scale into the conventional magnitude scale were available, it would be very useful in all research fields in seismology, since the magnitude scale has been widely used for nearly 40 years, not only for basic seismology but for applied, or engineering seismology. A new magnitude scale $M_w$ has recently been proposed by Kanamori (1977) to extend the usefulness of the old magnitude scale to large earthquakes. First, we must relate the seismic moment $M_o$ to the energy released in an earthquake. The reduction in the strain energy in an event may be expressed as the work done at the fault surface, $\Delta W = \sigma U A$, where $\sigma = (\sigma_1 + \sigma_2)/2$, and where the average offset, area, and stress before and after a slip at the fault are denoted by $U$, $A$, $\sigma_1$ and $\sigma_2$, respectively. Since $M_o = \mu U A$, the work done may be rewritten as $\Delta W = (\sigma/\mu)M_o$. If we assume that the stress after slip is equal to the frictional stress at the fault surface $\sigma_f$, i.e. $\sigma_2 = \sigma_f$, then the above equation reduces to

$$\Delta W - \sigma_f U A = (\Delta\sigma/2\mu)M_o, \qquad (6.5)$$

For most large earthquakes, $\Delta\sigma \approx 30$ bar and $(\Delta\sigma/2\mu)$ is approximately $1/(2 \times 10^4)$, suggesting the available seismic energy $E$ to be about $M_o/(2 \times 10^4)$. Taking this relation and the conventional energy–magnitude relation ($\log E = 1.5M + 11.8$) into consideration, Kanamori proposed the magnitude scale:

$$M_w = (\log M_o/1.5) - 10.7. \qquad (6.6)$$

Using this definition, the magnitudes of the Alaskan and Sanriku events, discussed above, are $M_w = 9.2$ and $M_w = 8.4$ respectively (see appendix 1 for several more examples).

### 6.2  FRACTURE OF ROCK

#### 6.2.1  *Failure problems in seismology*

Failures in rocks occur in various modes, depending on the properties of the rocks as well as on the environmental conditions. There are three fundamental types: *extension fracture, faulting* and *uniform flow*, which can be recognized from their macroscopic features (Griggs & Handin,

1966). Studies of failure were initially developed in materials engineering, where the strengths of materials are of primary concern. Rupture criteria have been discussed in connection with this problem in terms of the elastic and anelastic behaviour of a crack (or cracks), which is the hypothetical fracture element in a medium.

Successful application of the information yielded from this research field to seismic source mechanism studies has been achieved relatively recently with the development of theoretical and experimental work on rock properties under plausible conditions in the Earth. This work may be divided into two types of study. One is the application of crack theories in the study of source mechanisms enabling us to understand physical source parameters by analogy with crack models, as seen in §5.3.2 (see also §6.2.2). The second type is more concerned with laboratory testing of rock specimens, and experimental simulation of seismic fracturing on a reduced scale (see §§6.3 and 6.4).

### 6.2.2   Rupture criteria

*Macroscopic considerations.* Fig. 6.4a compares two types of stress–strain curves for rocks under failure test conditions. The abscissa represents the strain $\varepsilon$, and the ordinate, the *differential stress* $\sigma_1 - \sigma_3$, i.e. the difference of the maximum and minimum *principal stresses* (compressional) across the specimen (the intermediate principal stress is denoted by $\sigma_2$). The stress–strain curve AB is linear to a certain level of stress, B, beyond which abrupt fracture occurs (*brittle fracture*). The *strength* of the specimen is given by the level of the differential stress critical to fracture. The curve ACD represents *ductile* behaviour, in which stressing beyond a critical level C (*yield point*) leads to a marked increase of strain in the specimen with little increase in incremental stress. Fracture does not occur immediately when the yield point is reached.

In general, the strength of a brittle specimen increases as we apply *confining stress* to it (see fig. 6.12). The causal mechanism is not fully understood, but is conventionally interpreted in terms of the *Coulomb–Mohr rupture criterion*, as follows. Consider a two-dimensional stress field (fig. 6.4b), where $\sigma_n$ and $\tau$ denote respectively the normal and tangential stresses to a plane which makes an angle $\theta$ with the direction of the maximum principal stress $\sigma_1$. Then we have the relations: $\sigma_n = \frac{1}{2}(\sigma_1 + \sigma_3) - \frac{1}{2}(\sigma_1 - \sigma_3) \cos 2\theta$ and $\tau = \frac{1}{2}(\sigma_1 - \sigma_3) \sin 2\theta$ which imply that $\sigma_n$ and $\tau$ for a given $\theta$ may be graphically located on a circle in the $\sigma - \tau$ diagram in fig. 6.4c. The circle is centred at a point $\sigma = \frac{1}{2}(\sigma_1 + \sigma_3)$ on the $\sigma$-axis, and has a radius $\frac{1}{2}(\sigma_1 - \sigma_3)$. If a set of $\sigma$ and $\tau$ associated with fracture are yielded by a laboratory test, a circle (Mohr's circle) is specified on this diagram. A family of Mohr's circles can be drawn

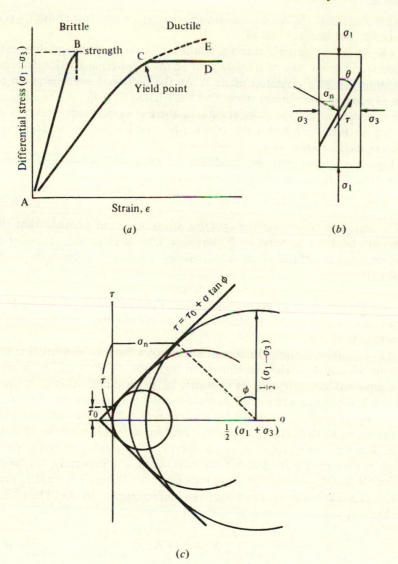

Fig. 6.4 (*a*) Brittle and ductile behaviour of rocks; (*b*) stress components in a rock specimen; (*c*) Mohr's circles and the rupture criterion.

by repeating failure tests under various conditions, as shown in fig. 6.4*c*. Coulomb noted that the envelope of the Mohr's circles can be well approximated, in many cases, by:

$$\tau = \tau_0 + \mu_i \sigma, \quad \text{or} \quad \tau = \tau_0 + \sigma \tan \phi, \tag{6.7a}$$

which represents the critical condition of $(\sigma, \tau)$ for brittle fracture ($\mu_i$ is a constant and denotes $\tan \phi$).

The strength increase due to confining stress may be qualitatively understood from this relationship, in terms of internal frictional stress resisting slip at a potential plane $\theta$. As the frictional stress is given as the product of the normal stress and the frictional coefficient, the higher the confining stress, the more resistance there is against slip. $\mu_i$ ($= \tan \phi$) and $\phi$ in (6.7a) are called the *internal friction (coefficient)* and *internal friction angle*, respectively.

Equation (6.7a) may be modified to take into account the pore pressure, $P$, due to water or fluid in rocks, as

$$\tau = \tau_0 + \mu_i(\sigma - P). \tag{6.7b}$$

This equation represents the *effective stress law*, and predicts that the strength tends to decrease as $P$ increases. This mechanism is thought to be important for the study of dilatancy models (wet-type) in rocks (§6.3.2).

*Griffith theory and modifications.* Further study of strength of materials leads us to a discussion of microscopic mechanisms of fracturing.

Let us follow Griffith (1920) and introduce a fracture element, a crack, in an idealized model as shown in fig. 6.5a. The crack is a two-dimensional free surface strip of width $2C$ in an infinite elastic medium which is subject to a tension $P$ in the direction perpendicular to the crack axis. Suppose the crack increases its width by an amount $dC$, then the work done in the system by the external force $P$ will increase by $dW_{\text{ext}}$ together with an increase in strain energy, $dW_{\text{el}}$, produced by the stress and strain concentration about the new edges. Consequently, the difference $dW_e = dW_{\text{ext}} - dW_{\text{el}}$ represents the energy available to drive the crack. The crack extension also increases the surface energy by $d\gamma_s$. The energy balance is therefore

$$\frac{\partial}{\partial C}(W_e - \gamma_s) = K, \tag{6.8}$$

and the crack will be stable if $K < 0$, and unstable if $K \geqq 0$. In the latter case, the crack will propagate spontaneously.

In the case shown in fig. 6.5a, $W_e$ and $\gamma_s$ are given by

$$W_e = \pi C^2 P^2 / E, \quad \gamma_s = 4CT, \tag{6.9}$$

where $E$ and $T$ denote the Young's modulus and the specific surface energy of the medium, respectively. Substituting (6.9) into (6.8) yields

$$\frac{\partial}{\partial C}\left(\frac{\pi C^2 P^2}{E} - 4CT\right) = K,$$

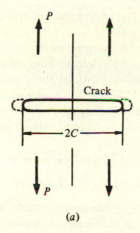

(a)

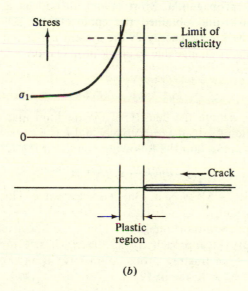

(b)

Fig. 6.5. (a) A crack under tension $P$; (b) the stress field at the edge of a propagating crack.

and                                                                    (6.10)

$$P_c = (2ET/\pi C)^{1/2},$$

which is the critical stress for $K = 0$. This indicates that the critical stress decreases as the crack length increases.

The Griffith's criterion, discussed above, has been supported by many laboratory tests. We may accept it as a basic rule. Further improvements of this criterion have been attempted by several workers (see Liebowitz,

1968, for review papers). The two problems that are of most concern are: anelastic behaviour due to stress concentration at the edges; and further application of the theory to dynamic cases.

Fig. 6.5*b* shows a stress distribution along the axis of a tensile crack, demonstrating the stress singularity at the crack tip. Microscopically, a certain region in front of the tip must yield plastically, leaving a thin layer of plastic deformation over the crack surface. Orowan (1950) length. In this case, the work associated with plasticity $\gamma_p$ is equal to $4CH_p$ so that (6.10) becomes

approximated the energy for this plastic work as a constant $2H_p$ per unit

$$P_c = [2E(T + H_p)/\pi C]^{1/2}. \tag{6.11}$$

That is to say, local plasticity can be taken into account by replacing the specific surface energy $T$ in (6.10) by an apparent energy $T' = T + H_p$.

Kinetic energy in the system must be considered to obtain a dynamic model of crack propagation. Mott (1948) introduced a kinetic energy term $\gamma_k$ into (6.8) and obtained the condition for self-exciting crack propagation. The kinetic energy in the system is derived from the time rate of strain-energy change due to crack propagation, and is given in the present model as

$$\gamma_k = \tfrac{1}{2} k \rho v_r^2 C^2 (P/E)^2, \tag{6.12}$$

where $\rho$ and $v_r$ denote the density of the medium and the velocity of crack propagation $(= dC/dt)$ respectively and $k$ is a constant. Adding the term $\gamma_k$ to (6.8) we obtain the following condition for $K = 0$

$$v_r = B\beta[1 - (\Delta\gamma_s/\Delta W_e)]^{1/2}, \tag{6.13}$$

where $\Delta W_e = \Delta W_{ext} - \Delta W_{el}$ (see p. 136), $\beta$ is shear wave velocity $(= (E/\rho)^{1/2})$ and $B$ is a constant.

$\Delta\gamma_s$ and $\Delta W_e$ are proportional to $C$ and $C^2$, respectively. This means that $v_r$ in (6.13) will approach $B\beta$ as the crack size increases. Several advanced models of fracture propagation have been proposed (see for example Takeuchi & Kikuchi, 1973).

### 6.2.3 Friction and stresses in a fault

*Stress changes in faulting.* The discussion above has shown that stress concentration at the crack tip plays a critical role in crack propagation. If an observer is located at a point lying on the outer extension of the crack axis and the crack propagates toward it, then the stress change which appears at the point will be similar to the curve shown in fig. 6.5. The stress increases quickly, as the front passes, from the initial level up to a high level, and then falls to zero when the point enters the free surface region.

Stress changes in propagating shear fractures will be similar to this basic trend. In faulting, however, the stress is closely related to friction in the fracture, which will make the stress-change curve more complicated. Fig. 6.6 presents a schematic view of stress change in faulting modified after Yamashita (1976), where

$\sigma$    stress, in a general sense (e.g. tangential stress),
$\sigma_1$    initial stress,
$\sigma_2$    final stress,
$\sigma_f$    frictional stress (dynamic),
$\sigma_{fr}$    frictional stress (static).

These are the elementary components. *Frictional stresses* are generally determined by the surface conditions and the normal stress on the fracture surface, so that they may be regarded as constants. If the fault has a considerable depth, of course, the normal stress and frictional stress will vary with depth. Other parameters in the figure are deduced from the above four quantities. For example, $\sigma_{frs}$ is the difference between the frictional stress and the initial stress, and indicates the minimum increment of the initial stress for rupturing. If the local stress concentration exceeds this value, a slip occurs in that section of fault. $\sigma_{eff}$, which is called the *effective stress*, denotes the difference between static and dynamic friction. This term is sometimes used to denote $\sigma_1 - \sigma_f$ instead of $\sigma_{fr} - \sigma_f$ (e.g. Brune, 1970; his model assumes instantaneous propagation, so that $\sigma_{frs} = 0$ and $\sigma_1 = \sigma_{fr}$, therefore $\sigma_{eff}$ is equal to $\sigma_1 - \sigma_f$). The stress in the faulted portion might not remain at $\sigma_f$ for long if it is subject to stress and strain readjustment by further fault propagation to adjacent portions. The residual stress $\sigma_2$ is shown at a different level from $\sigma_f$, making the case general. The *stress-drop*, $\Delta\sigma$, as given from long-period waves,

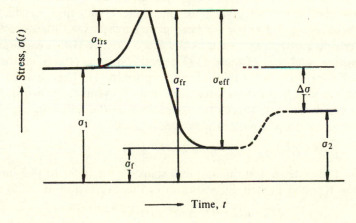

Fig. 6.6. Stress change with time on the surface of a propagating shear fracture. (After Yamashita, 1976.)

denotes the difference between $\sigma_1$ and $\sigma_2$. The ratio of stress-drop to the effective stress available for fracturing, $\varepsilon = \Delta\sigma/\sigma_{\text{eff}}$, is sometimes called the *fractional stress-drop*.

The stress-drop and effective stress can be derived from seismic data, but the absolute values of the stresses cannot. It has been suggested that the fault slippage will produce an energy $\sigma_f U$ per unit area (where $U$ is the slip amplitude), which will be dissipated as heat. Using this principle, Brune, Henyey & Roy (1969) studied heat-flow data in the San Andreas fault and concluded that the upper limit of $\sigma_f$ must be of the order of a few hundred bars. If local melting of rocks takes place on a fault surface, however, the friction will drop to a very low value as Brune (1970) suggested.

In the discussion above, we considered the static friction and dynamic friction as constants, for the sake of simplicity. This condition, however, may not always hold. Scholz, Molnar & Johnson (1972) considered $\sigma_f$ as the average frictional stress and expressed $\Delta\sigma$ as

$$\Delta\sigma = 2[\sigma_1 - \sigma_f/(1-\eta)], \tag{6.14}$$

where $\eta$ is the seismic radiation efficiency, discussed later in this section. Notice that $\Delta\sigma$ depends on $\eta$, even if $\sigma_1$ and $\sigma_f$ are fixed. We may assume that $\eta$ is much less than 1 in practice and with this assumption we obtain:

$\Delta\sigma \leqq \sigma_1 - \sigma_f$  if $\sigma_1 \leqq (1 + 2\eta)\sigma_f$ (Brune, 1970);
$\Delta\sigma > \sigma_1 - \sigma_f$  if $\sigma_1 > (1 + 2\eta)\sigma_f$ (Savage & Wood, 1971);
$\Delta\sigma = \sigma_1 - \sigma_f$  if $\eta = 0$ (Byerlee, 1970*a*).

*Stress-drops in large earthquakes.* The stress-drops, $\Delta\sigma$, derived from seismic wave data fall mostly in a range 10–100 bar. Kanamori & Anderson (1975) re-analysed the data and concluded that 30 bar was a typical value for *inter-plate earthquakes* (those occurring along the major plate boundaries such as the Chile (1960) and Alaska (1964) earthquakes), and 100 bar for *intra-plate earthquakes* (other crustal and sub-crustal earthquakes, such as the Kern County (1952) earthquake in California and the Tango (1927) and Niigata (1964) earthquakes, Honshu, Japan – see table 7.3 and appendix 1). They compared the seismic moment and the source area in each event, as shown in fig. 6.7, from which we may deduce the stress-drop in the following way.

The stress-drop is written in a general form as,

$$\Delta\sigma = C\mu(U/A^{1/2})$$

where $C$ denotes a non-dimensional shape factor (refer to (4.7) and (4.8) for the Knopoff model). Substitution of this expression in (4.20) gives

$$M_0 = \mu U A = \tfrac{16}{7}\Delta\sigma a^3 = (16\Delta\sigma/7\pi^{3/2})A^{3/2}, \tag{6.15}$$

where $C$ is evaluated for a circular crack of radius $a$ (Eshelby, 1957). This

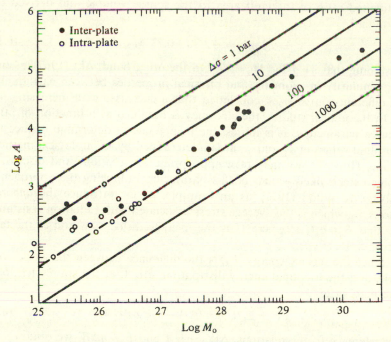

Fig. 6.7. Relation between the fault surface area, $A$ (km$^2$), and the seismic moment $M_0$ (dyn cm). The straight lines give the relations for circular cracks with constant stress-drop ($\Delta\sigma$). (After Kanamori & Anderson, 1975.)

suggests a linear relation, $\log A \approx \frac{2}{3} \log M_0$, and enables us to draw a family of equi-$\Delta\sigma$ lines, as shown in fig. 6.7.

The experimental data in the figure lie between 10 and 100 bar. The values for inter- and intra-plate earthquakes are clearly different. The relatively low stress-drop in the former group reflects frictional conditions in the subduction zones and transform faults fundamentally different from those in a plate itself. It is rather surprising that seismic events of either type are related to a constant stress-drop over a wide range of moment and magnitude. This gives us the impression that fault slips in large earthquakes are not very sensitive to local conditions, as far as the stress-drop is concerned.

*Seismic efficiency.* Seismic efficiency, $\eta$, is the ratio of the radiated seismic wave energy, $E_S$, to the change in strain energy at the source, $\Delta W$, i.e.,

$$\eta = E_S/\Delta W. \qquad (6.16)$$

Båth & Duda (1964) assumed that $\eta$ is proportional to an empirically estimated ratio of the fault area to the source volume and derived the formula

$$\log \eta = -2.00 + 0.23 \, M_S, \tag{6.17}$$

assuming that $\eta = 1$ for $M_S = 8.7$. On the other hand, Aki (1967), assuming similarity in geometry and physical properties between earthquakes of different source sizes, concluded that $\eta$ decreases with increasing $M_S$.

In theoretical studies, the efficiency is obtained as a function of other source parameters. It is impossible in practice to determine $\eta$ since the absolute values of $\sigma_1$ and $\sigma_2$ are undetermined by seismological methods alone. However, we may take appropriate assumptions and discuss, to some extent, likely values of this parameter. Following the approach of Savage & Wood (1971), as an example, let us introduce the *apparent stress*, $\sigma_a$, which is the average stress associated with radiation resistance, so that $E_S = \sigma_a U A$ (where $U$ is the average dislocation and $A$ the fault area, cf. (5.28)).

Since the seismic energy, $E_S$, is the difference between the work done, $\Delta W$, and the frictional energy dissipation, $\sigma_f U A$, at the source (cf. (6.5)), we may write

$$\sigma_a = E_S/UA = (\Delta W - \sigma_f U A)/U A. \tag{6.18}$$

In reference to the relations, $\Delta W = \sigma U A$ and $E_S = \eta \Delta W$, we obtain

$$\eta = \sigma_a/\sigma = (\sigma - \sigma_f)/\sigma$$

or

$$\sigma_f = \sigma - \sigma_a, \tag{6.19}$$

where $\sigma = (\sigma_1 + \sigma_2)/2$. Savage & Wood supposed overshooting of the fault, associated with the inertia of the moving fault blocks. In other words, the slipped fault locks again at a stress $\sigma_2$ somewhat less than the dynamic-friction stress $\sigma_f$, i.e. $\sigma_f > \sigma_2$. In this case, we derive from (6.19) the following stress relation:

$$\sigma_a < \tfrac{1}{2}\Delta\sigma, \tag{6.21}$$

which is called the Savage–Wood inequality.

If $\sigma_a \approx \tfrac{1}{2}\Delta\sigma$, and consequently $\sigma_f \sim \sigma_2$, as Kanamori & Anderson preferred, we may approximate $\eta$ as

$$\eta = (\sigma - \sigma_2)/\sigma = \varepsilon_0/(2 - \varepsilon_0),$$

where $\varepsilon_0 = \Delta\sigma/\sigma_1$. If the stress-drop is unchanged, this suggests such a trend that the higher the initial stress, the lower the seismic efficiency.

Kanamori & Anderson (1975) studied empirical formulae relating various source parameters, and discussed the efficiency in relation to the initial tectonic stress, $\sigma$. For example, they showed $\eta = 1.0$, 0.43, 0.11, and 0.026 for $\sigma = 60$, 100, 300, and 1200 bar respectively (assuming a stress-drop of 60 bar), although they did not express a preference for any particular value.

### 6.3 DILATANCY AND STICK-SLIP

#### 6.3.1 *Experimental approaches*

A beautiful demonstration of a dilatancy process may be seen on the beach. As one stands on wet sand, the sand around the foot seems to lose water saturation for a short time. Pressing the beach surface disturbs the close compaction of sand grains, thus causing momentary undersaturation of water due to a local porosity increase. *Dilatancy* is this non-elastic volume increase caused by stress, and has been recognized since early this century in soil mechanics studies. Bridgman (1949), and more recently Brace, Pauling & Scholz (1966), noted similar volume changes in rock specimens, and attributed them to the formation of numbers of small cracks in the specimen.

Laboratory experiments (Brace, 1968; Scholz, 1968) have clarified the nature of dilatancy in rock samples. A general pattern emerges as shown in fig. 6.8. When compressive stress in the sample reaches about half its fracture strength (vertical broken line in fig. 6.8), cracks begin to appear with planes tending to be parallel to the plane of the greatest compressive strength. Microcrack grow as stress increases, then coalesce to form one or more principal fractures in the final fracture stage. The bottom trace in fig. 6.8 is the number of acoustic emissions (i.e. high-frequency sound pulses due to micro-fracturing) from cracking, which increases enormously with stress above 95% of the fracture stress.

The $\Delta V/V_0$ trace (third from the top) in the figure shows volume increase in the dilatancy process. It is very similar to the graph of micro-fracturing, suggesting the increase in the number of pores is an immediate cause of the volume increase. A value of 0.3% was attained at 100% fracture stress in this example. The development of cracks will cause variations to occur in other physical properties of the medium with increasing stress. For example, we notice significant changes in wave velocity (top curves in fig. 6.8) and electrical resistivity (second curve in fig. 6.8), although the variations appear to be quite complicated. Note that the dilatancy effect on wave velocity is anisotropic, predominating in the direction perpendicular to the compressive stress axis. The velocity measured in the parallel direction is insensitive to the dilatancy effect. This is physically understandable, because the arrangement of fracture planes parallel to the plane of the least compressive strength (as

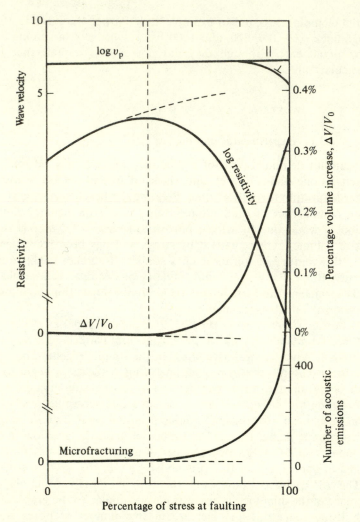

Fig. 6.8. Changes in the physical properties of granite under compressive stress. Tests were conducted on Westerly granite under several kilobars confining pressure. (Volume, resistivity and wave velocity data are from Brace, 1968; microfracturing data are from Scholz, 1968. The wave velocities parallel (∥) and perpendicular (⊥) to the compressive stress axis are shown.)

mentioned above) would affect wave propagation more seriously in the perpendicular direction than in the parallel direction.

### 6.3.2   Dilatancy models

The discovery of premonitory changes in seismic velocities and other physical properties prior to earthquakes (see chapter 8) has urged seismologists to establish physical bases for earthquake prediction using this kind of information. The concept of dilatancy and experimental data of the type described above are very useful for this purpose. Basically, two different kinds of model are used. These are the *wet* and *dry* models, depending on whether the model takes the role of water (or fluid) flow into consideration or not.

*Wet model.* The experimental data in fig. 6.8 allow us to draw a preliminary picture of the premonitory velocity changes if we hypothesize the occurrence of dilatancy in a seismic area. However, this idea on its own is not sufficient to explain the mechanism which causes recovery of the diminished velocities back to their former level, as observed in the second half of the precursory period (see chapter 8). A persuasive explanation for this effect was suggested by Nur (1972), who showed that seismic velocity in a dilatant medium might depend strongly on whether the cracks were dry or wet by water (or fluid). In general, the velocity in the dry dilatant medium is lower than that in a water-saturated medium, so the mechanism of velocity recovery could be explained by water diffusion in the medium. Scholz, Sykes & Aggarwal (1973) took a similar idea and proposed a more generalized model (fig. 6.9). This type of model is sometimes called the *dilatancy–diffusion* (*DD*) *model*, after Anderson & Whitcomb (1973).

Fig. 6.9 is a qualitative picture of the predicted temporal changes of various physical properties for the six stages in a dilatancy seismic cycle. Let us assume the crustal stress in the medium increases with time, presumably at a uniform rate. The stress (and strain) builds up through stage I to a certain critical level (fig. 6.8 suggests 50% of the ultimate value), when dilatancy starts to predominate. In stage II, dilatancy cracks develop causing undersaturation of the medium and decrease of velocities. From laboratory experiments, this effect appears in $v_P$ more than in $v_S$, resulting in a decrease of the $v_P/v_S$ ratio by 10–20% of the normal value. Land uplift and tilt are also expected in this stage. Stage III is characterized by re-saturation with water which causes the decreased velocities to regain their normal values. During stages II and III, the frequency of seismic events should decrease, since undersaturation will lower the *pore pressure* in such a way that further fracturing will be suppressed (*dilatancy hardening*). However, this phenomenon will cease, sooner or later, since the influx of water from the surrounding medium will increase the pore pressure again and consequently lower the effective strength of the rocks. A catastrophic fracture will take place shortly after re-saturation. In this critical period, we might observe foreshocks.

The model allows us long- and short-range prediction in terms of dilatancy stages. Commencement of stage II will be recognized by characteristic changes in the velocity ratio and land movement, and also

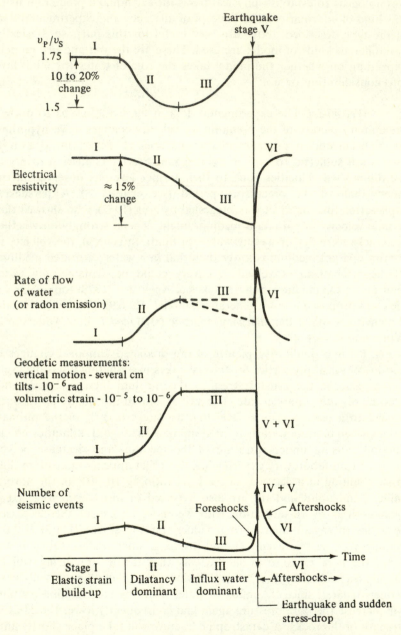

Fig. 6.9. Changes in physical parameters predicted by the dilatancy–diffusion model. Roman numerals indicate various stages in a seismic cycle. (After Scholz, Sykes & Aggarwal, 1973. Copyright 1973 by the American Association for the Advancement of Science.)

in other physical parameters such as electrical resistivity, rate of water flow and radon emission, as seen in fig. 6.9. If these individual components show anomalies simultaneously, we may be confident that stage II has been recognized. Recovery of the velocity ratio is a sign of the recovery of water saturation, and consequently of the irreversible process toward the final catastrophe. This effect is useful for short-range predictions. Scholz, Sykes & Aggarwal (1973) compiled the available data on the duration time of premonitory events in various cases and proposed the following empirical formula

$$\log T = 0.685 \, M - 1.57, \tag{6.22}$$

where $T$ is the precursor time in days, which they attributed to the delay time in pore pressure change by water diffusion through a finite volume of a dilatant medium (see also §8.3.2).

*Dry model.* Basic features of seismic precursors may also be explained by a dry model. Models by Brady (1974), Mogi (1974) and Stuart (1974) are of this type, in which the inhomogeneous development of dilatancy cracks plays a principal role, but where the hypothesis of water diffusion is not required. The dry model was originally developed by Soviet seismologists (see Mjachkin *et al.*, 1975) and was called the *IPE model*, in reference to the abbreviation of the Institute of Physics of the Earth where the model was constructed. The basic concepts of the IPE model were summarized by Mjachkin *et al.* as follows:

1. Fracture of statistically heterogeneous materials is caused by the increase of number and size of crack-like defects.
2. The defects may develop in time under approximately constant stress and the rate of their formation increases with the increase of stress.
3. The total deformation consists of intrinsic elastic deformation and deformation caused by mutual displacement of crack edges.
4. Macrofracture (development of the main fault) is the result of the avalanche-like growth and resulting instability which occurs when a certain density of cracks is reached.
5. Formation of the main fault results in the decrease of stress level in the surrounding volume; as a result, the growth of new defects stops and number of active cracks is decreased.
6. The fracture process does not depend strongly on scale.

According to the IPE model, the medium should be divided into two zones, A and B, where A is the zone of future faulting (unstable deformation) and B is the surrounding zone of future unloading. The development of processes prior to the earthquake is described in terms of four successive stages, I–IV, as shown in fig. 6.10.

Suppose a volume of statistically homogeneous medium (rocks) is subject to stress increasing with time. In stage I, cracking is statistically homogeneous in zones A and B, and there is a gradual increase in crack

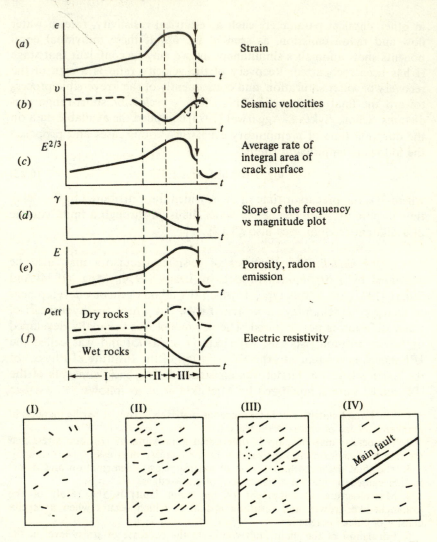

Fig. 6.10. Changes in the physical parameters predicted by the IPE model. The four principal stages in a seismic cycle with respect to fracture development are illustrated at the bottom of the figure. (After Mjachkin *et al.*, 1975.)

density with time. When the crack density exceeds a certain critical level, mutual interaction of cracks accelerates their development, causing an avalanche in stage II. As a result, we will observe a rapid increase in strain rate and porosity and a rapid decrease in seismic wave velocities (see fig. 6.10). This process will result in unstable cracking in stage III, i.e. further increase of deformation will form relatively large cracks in a narrow zone A, and consequently the decrease of stress and strain will appear in the surrounding zone B. After this has occurred the main fractures in zone A develop at an increasing rate until the fractures finally coalesce into a catastrophic fault (stage IV).

Fig. 6.10 illustrates temporal changes in several parameters predicted by the IPE model for zone A. Notice that basic features of seismic ᵖrecursors, as discussed in the previous wet model, can be explained by ᵗᵉ dry model as well. However, the IPE model predicts two alternative possibilities for the electric resistivity according to whether the medium is dry or wet (see the bottom curve in fig. 6.10).

To summarize, the two models explain the observed seismic precursors equally well, but are based upon fundamentally different physical processes. Mjachkin *et al.* (1975) compared the two models and suggested some diagnostic tests by measurement of stress near the focal region. If the IPE model is valid, significant stress-drop should be observed prior to the main fracture. Measurement of pore pressure will also be a useful test for the two models, since its change in time is an essential factor in the dilatancy–diffusion model. There are several other possible methods of testing the validity of the two models, but it is difficult, at the present time, to decide which model should be adopted.

### 6.3.3  *Stick-slip*

Brace & Byerlee (1966) showed that previously faulted surfaces of rock may undergo a series of intermittent small slips under continuing pressure, which they called *stick-slip* in reference to a term used in materials science, and suggested that this mechanism might apply to earthquakes (fig. 6.11). Processes in stick-slip are described by Byerlee (1970*b*) as follows:

When two surfaces of a brittle material are placed together, the asperities on the surfaces in contact become locked together. If the normal load is high enough to prevent the surfaces from lifting up over the irregularities, then sliding will occur when the locked regions fail brittlely.

Macroscopically, stick-slip is an unstable type of frictional sliding of two blocks in contact, in which the differences of static and kinetic (sliding) frictions at the contact surfaces play an essential role. Let $\mu_s$, $\mu_k$ and $\sigma_n$ denote, in an average sense, the coefficient of static friction, coefficient of kinetic friction and normal stress to the surfaces, respectively. Then the stress-drop $\Delta\sigma$ in the sliding is given as

$$\Delta\sigma = 2[\mu_s - \mu_k/(1-\eta)]\sigma_n \qquad (6.23)$$

by Scholz, Molnar & Johnson (1972), where $\eta$ is the seismic efficiency (§6.2.3).

Suppose the shear stress across the contact surfaces increases, then the

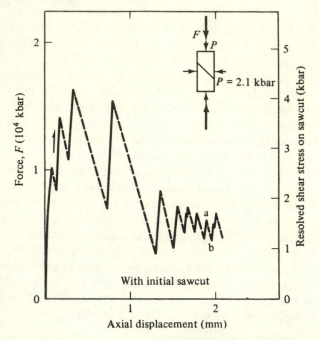

Fig. 6.11. Force–displacement curve illustrating stick-slip on sawcut of Westerly granite at 2.1 kbar confining pressure. (From Brace & Byerlee, 1966. Copyright 1966 by the American Association for the Advancement of Science.)

blocks start to slide when the stress exceeds the static friction $\mu_s\sigma_n$, causing a stress-drop given by (6.23). If $\eta = 0$, i.e. the case of no seismic wave radiation, (6.23) is reduced to $\Delta\sigma = 2(\mu_s - \mu_k)\sigma_n$, which suggests that stress will vary between $\mu_s\sigma_n$ and $(2\mu_k - \mu_s)\sigma_n$.

Occurrence of stick-slip depends in a complicated way upon the mechanical properties (e.g. surface finish, mineral composition etc.) and hysteresis of the contact surfaces and the physical environments of the blocks. Laboratory tests show that sliding of the blocks tends to turn from stable sliding to stick-slip, with increasing normal stress (or confining pressure), at room temperature. Fig. 6.11 is taken from Brace & Byerlee (1966), and illustrates stick-slip in Westerly granite test blocks at 2.1 kbar confining pressure.

The concept of stick-slip is undoubtedly an important clue for understanding of rock-mechanics in shallow earthquakes. At the moment, however, little is known about the mechanical properties and stress environments of fault surfaces and gouges. Therefore, it is imperative that field data, as well as data from laboratory experiments on the mechanical and environmental conditions, should be accumulated, so that we may reach a better understanding of the role of this process in shallow earthquakes.

## 6.4 PHYSICS OF DEEP-FOCUS EARTHQUAKES

### 6.4.1 *Elementary considerations*

Two further difficulties must be faced when deep-focus earthquakes are studied.

The first arises from the paucity of field data. It is impossible, in practice, to use near-field data, or even surface wave data, so that seismic body waves are the only sources of information on movement at a deep seismic origin. As far as the radiation patterns of P- and S-waves are concerned, however, the focal mechanism of deep and intermediate earthquakes is similar to that of shallower ones (see for example Honda, 1957). Therefore, it is reasonable to attribute a deep-focus earthquake to a certain type of shear failure, though it might not be exactly like the brittle fracture supposed for shallower earthquakes (see Richter, 1958, for a historical review).

The second problem concerns the physical conditions at a great depth, where the following effects would appear due to the high confining pressure:

    (*a*) extremely high frictional resistance to slip;
    (*b*) brittle–ductile transition;
    (*c*) relaxation of differential stress, if plastic flow occurs.

As pointed out earlier by Jeffreys (1936), and studied later by Griggs & Handin (1960) and Orowan (1960) in detail, frictional resistance to slip increases nearly proportionally to pressure (and consequently depth), so that it may exceed the yield point or the strength of rocks beyond a certain depth. In other words, *frictional faulting*, if it occurs, must be driven by extremely high shear stress against friction. At the depth of 600 km, for example, the shear stress level, and consequently the strength of the rock, would need to be approximately 400 kbar, which is an unbelievably high value.

The *brittle–ductile transition* will occur in rocks if the confining pressure exceeds a certain limit. Fig. 6.12 illllustrates compressive strength versus confining pressure for silicate rocks. Under low confining pressure, rocks undergo mostly brittle failure as indicated by open symbols in the figure. As the pressure increases, however, ductile behaviour becomes predominant, especially in rocks of relatively low strength, and consequently we may divide the diagram into the brittle and ductile domains.

These data suggest that instability mechanisms, such as frictional faulting in shallow earthquakes, do not apply immediately to deep-focus earthquakes. A number of different hypotheses have been proposed for the instability mechanism of a deep seismic source. These theories may be categorized into two broad groups, depending on whether they favour

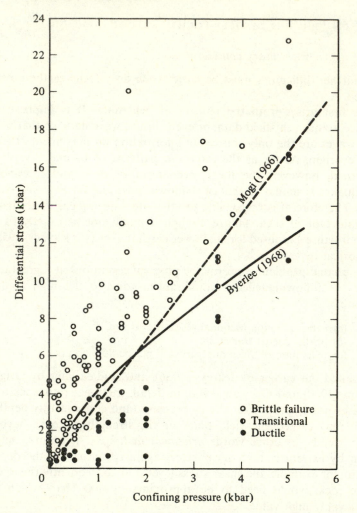

Fig. 6.12. Compressive strength (differential stress) versus confining pressure for silicate rocks (data from Brace, 1969). The boundaries between brittle and ductile behaviour based on friction data prepared by Mogi (1966) and Byerlee (1968) are represented by the dashed and solid curves, respectively. (From Dieterich, 1974. Reproduced, with permission, from the *Annual Review of Earth & Planetary Sciences*, Volume 2. © 1974 by Annual Reviews Inc.)

shear instability or not. The first group, favouring shear instability, may be divided into two further subgroups depending on whether brittle or ductile fracture is assumed. §§6.4.2, 6.4.3 and 6.4.4 discuss several of these theories in more detail.

### 6.4.2    *Brittle fracture under pressure*

*Stick-slip.* Stick-slip was originally recognized as a principal instability mechanism for shallow earthquakes (§6.3.3). Later, Byerlee &

Brace (1969) widened its application to deep earthquakes, on the basis of their laboratory experiments under higher pressure (fig. 6.11). In fact, they observed stick-slip behaviour on previously faulted surfaces under normal stresses as high as 17 kbar. Snapping phenomena resembling stick-slip might also occur under even higher pressures, but it is questionable whether these may always be attributed to dry frictional instability. Even if we accept the possibility of superficial stick-slip facilitated by water, its application would be limited to a depth of 100 km, beyond which dehydration of minerals is likely to be complete (Griggs & Baker, 1969).

*Dehydration.* Experiments by Raleigh & Paterson (1965) confirmed that fluid pressure acts to extend the range of brittle behaviour to higher pressure in serpentinites (fig. 6.13). This effect is explained fundamentally by *dehydration* of the serpentinite. In fact, ductile behaviour is observed at relatively low temperatures, but brittle behaviour, indicated by rapid stress-drop, commences with the onset of dehydration at higher

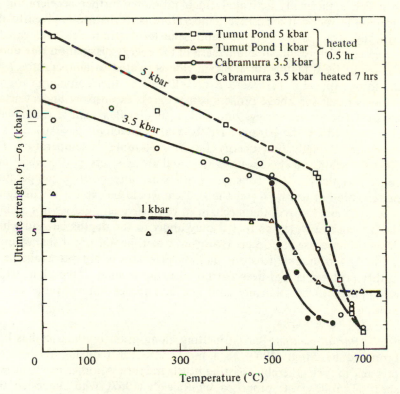

Fig. 6.13. Ultimate strength of antigorite-chrysotile serpentinites from two locations at different temperatures and pressures showing the drop in strength arising from dehydration. (After Raleigh & Paterson, 1965.)

temperatures. Several theories have been proposed for its mechanism, they are:

> (*a*) dehydration causes pore pressure to increase and effective confining pressure to decrease, consequently friction at the grain boundery is lowered to cause slipping;
> (*b*) water from dehydration lubricates surfaces of previous faults (or Griffith cracks) to trigger slipping;
> (*c*) if dehydration takes place in Griffith cracks, volume increase of the cracks causes local tension at their edges and triggers brittle fracture.

In any case, its application to a deep seismic source poses fundamental geophysical questions; can hydrous minerals exist to depths as great as 700 km, and does dehydration occur there?

### 6.4.3    *Ductile fracture under pressure*

*Shear melting.* Griggs & Baker (1969) studied thermodynamical conditions at a great depth and proposed the *shear melting* mechanism, in which a thermally activated creep promotes further acceleration of creeping in rocks. If a positive temperature perturbation is given to the critically strained medium at depth, the local strain rate will rapidly increase, which results in an increase of the energy dissipation rate and a rise in the local temperature. Strain rate is related to temperature $T$ by the form $\exp(-E/RT)$, where $E$ and $R$ are the thermodynamical constants, so that the above process can progress explosively under certain physical conditions. Griggs and Baker simulated a mantle earthquake characterized by the parameters: thermal conductivity $k = 0.01$ cal s$^{-1}$ cm$^{-1}$ deg$^{-1}$; activation energy $E = 40$ kcal mole$^{-1}$; temperature $T = 1000$ K; strain rate $\dot{\gamma} = 3 \times 10^{-4}$ s$^{-1}$; and shear stress $\tau = 0.3$ kbar; and concluded that, in the constant stress case, any positive temperature perturbation of width greater than 30 km would be explosively unstable.

As they emphasized, the mechanism of shear melting does not assume the presence of water, so that it may apply to the depths beyond which dehydration ceases to occur. However, experimental proof that fracture is totally attributable to shear melting processes is not yet available. A further problem is that deep earthquakes are located in the plate, which is supposed to be relatively cool and is a rather unfavourable site for melting.

*Partial melting.* Partial melting along grain boundaries has been proposed by Mogi (1967) as a mechanism for reducing the effective pressure in rocks thereby assisting brittle fracture. Another mechanism of shear instability was proposed by Raleigh (1967), who suggested that partial melting tends to concentrate in pores and accelerate creep, causing enlargement of pores in a preferred direction.

Savage (1969) speculated that surfaces of one shallow fault are preserved as planes of weakness in a slab during its down-going motion, and that partial melting takes places to fill up large lenticular cavities thus brought into deep regions. If this is the case, slips in deep seismic sources will occur along planes of weakness already 'fixed' in the slab.

### 6.4.4 *Phase transitions*

Explosive phase transitions at depth have also been discussed by a number of seismologists, such as Benioff (1964), Evison (1963), Dennis & Walker (1965) and others. It is generally difficult to envisage such a mechanism, in which only rigidity changes occur, with density and compressibility remaining unchanged. Therefore, application of the phase-transition theory to deep seismic sources means that the basic concept of shear fracture must be abandoned.

This is in conflict with our seismological picture of a deep-focus earthquake (see §6.4.1; see also Sykes, 1968), and must be regarded as questionable unless significant deviation of radiation pattern from the quadrant type is observed. Several papers have reported data showing no quadrant-type of radiation pattern (e.g. Benioff, 1951; Randall & Knopoff, 1970). However, these data are too scarce and dispersed to persuade seismologists to altogether abandon the shear fracture picture of deep earthquakes.

Ringwood (1972), and later McGarr (1977), speculated that the voluminal contraction of rocks in a descending slab due to phase transition might develop large shear stresses in the material surrounding the contracting volume, resulting in earthquakes. McGarr compared the depth ranges of predominant activity of intermediate and deep-focus earthquakes in the upper mantle to the depths of possible phase changes and justified the speculation from the good correlation between the two kinds of data.

In summary, the mechanism of intermediate and deep earthquakes remains an open question, in spite of vigorous debate among seismologists. Progress in plate tectonics has renewed seismologists' interest in deep earthquakes. Thus, extensive amounts of observational data are being accumulated to clarify their mechanism in relation to the tectonic behaviour of down-going slabs. With precise information about deep seismic sources, it may be expected that our understanding of the nature of deep earthquakes will significantly progress in the next several years.

# 7    Earthquakes and tectonics

The seismic geography of the world shows that the great majority of earthquakes are concentrated in relatively narrow zones, i.e. *seismic zones* (§2.4.2). Some of the zones are shallow, but others extend to great depths forming *seismic planes*. Seismic and tectonic activity at these zones or planes are basically attributed to interactions of adjoining tectonic plates which undergo relative motion across their boundary surfaces (§7.1.3). In this respect, each zone (or plane) can be thought of as a huge macroscopic fault, i.e. a *megafault*.

This chapter discusses the tectonic implications of earthquakes, but not the Earth's tectonics as such. We may briefly discuss tectonic problems as occasion demands, but a detailed study of the individual tectonic zones as well as specific discussion of plate tectonics will be left to other references (see for example Bird & Isacks, 1972; Le Pichon, Francheteau & Bonnin, 1973; Sugimura & Uyeda, 1973; Lomnitz, 1974).

### 7.1.1    *Behaviour of the San Andreas fault*

The tectonic pattern in California is dominated by the northwest-trending San Andreas fault, which is traceable for 1100 km across land from the Gulf of California northwards to beyond Point Arena. The motion is essentially right-lateral, with the west block moving north relative to the east block. The geological features are complicated in detail by numbers of associated faults, striking parallel (e.g. Calaveras and Hayward in central California and San Jacinto and Imperial in the south) or transverse (e.g. Garlock) to the main fault (fig. 7.1).

This region is part of the circum-Pacific seismic zone, and has generated many large and small earthquakes, producing almost 30 major earthquakes in the last 150 years (shocks in Nevada are included). It is undoubtedly a region of primary importance both for seismology and other solid earth sciences and, as such, has been researched extensively by many different disciplines. We shall draw on this large body of literature in the following discussion. Let us review several basic aspects of the fault for future reference. Further information on the San Andreas

156

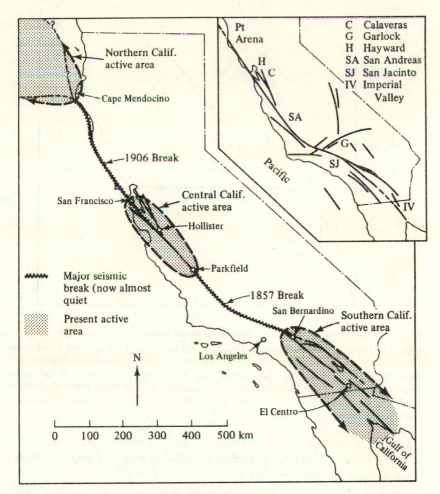

Fig. 7.1. Areas of contrasting seismic behaviour along the San Andreas fault zone in California. (After Allen, 1968.)

fault may be obtained from many papers and books (see for example Richter, 1958; Gutenberg & Richter, 1954; Dickinson & Grantz, 1968; Kovach & Nur, 1973; Lomnitz, 1974).

*Cumulative fault offsets since geological time.* Stratigraphic studies have discovered large cumulative right-lateral slip on the San Andreas fault system amounting to several hundreds of kilometres in the past tens of millions of years (Hill & Dibblee, 1953). Clarke & Nilsen (1973) combined the results of several geological and geomorphic studies to show the history of the fault offset in central California (fig. 7.2). It is remarkable that the fault accumulated right-lateral slip between about 80 and 60 million years ago, but that no appreciable offset accumulated for the

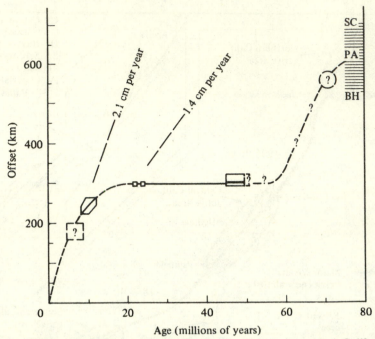

Fig. 7.2. History of offset along the San Andreas fault in central California. Probable limits on the age and offset distance of certain displaced features are indicated by the polygons. The shaded area at the right-hand side of the figure illustrates offsets in the Mesozoic granite basement terrain from Eagle Rest Peak to: BH – Bodega Head; PA – Point Arena; SC – Shelter Cove. (After Clarke & Nilsen, 1973.)

intermediate period (between about 60 and 20 million years ago), after which the present trend of movement started at a rate of about 2 cm per year, around 20 million years ago.

*Continuing fault offsets.* Geodetic work along the fault by serveral groups discovered coseismic movements in major earthquakes, which accumulated the right-lateral offsets in the respective segment of the fault. As well as incremental fault movements of this type, creep-type offsets are found in several special localities, as first observed by Tocher (1960) in the town of Hollister using the distortion of a winery building and the offset of other linear man-made features. Brown & Wallace (1968) studied this type of data and concluded that there is an average yearly offset of 2 cm on the segment of the fault about Hollister, central California (see fig. 7.1). A modern study of geodetic data by Savage & Burford (1973) concluded that the current offset rate in central California is about $3.2 \pm 0.5$ cm per year.

*Earthquakes on the fault.* Causal relations between plate tectonics and seismic activity are evident in many respects, i.e. alignment of epicentres of large and small earthquakes along fault traces as well as agreement of their focal mechanisms and fault breaks (if observed) with the mode of fault offset at each locality. Also notable is the absence of earthquakes below a certain depth (15–18 km), which characterizes seismic activity in California. This suggests that the deep fault surface is subject to special physical conditions, presumably plasticity, and does not accumulate strains effectively.

*Two contrasting types of strain release.* Fig. 7.1 divides the fault into segments and classifies them into two types distinguished by contrasting modes of strain release (Allen, 1968). The first type involves the segments of the 1857 and 1906 breaks which are the sites of repeated great earthquakes. During the interseismic period, however, no appreciable fault movements are observed in these segments. The second segment type is characterized by continuous creep or by numerous smaller earthquakes, like those in central and southern California. Great earthquakes do not seem to occur in these segments.

*Tectonic environments.* Macroscopically, activities of the San Andreas fault system represent northwestward drifting of coastal California with respect to the main mass of North America. The data in fig. 7.2 allow us to trace its history back to the early Miocene, although further extrapolation of the curve cannot be done uniquely.

Atwater (1970) studied sea-floor magnetic lineations off California and suggested a history for the growth of the San Andreas fault system. According to her model, plate subduction occurred off the coast until about 32 million years ago. When part of the East Pacific rise, which was moving northward relative to the North American plate, 'hit' California, strike-slip movement was initiated along the part of the plate boundary where the rise met the trench. The segment lengthened with time, and now the north end of the boundary is at Cape Mendocino and the south end in the Gulf of California (see fig. 7.1). The right-lateral slip along the San Andreas fault system, i.e. the northwestward movement of the sea-side block relative to the land-side one, may be accounted for by a combination of northward movement of the rise and westward spreading of the Pacific plate from it, where both movements are relative to the North American plate. The San Andreas fault system seems, in this respect, to be a relatively new transform fault (§7.1.3), which has developed obliquely to the pre-existing submarine fracture zones which trend east–west.

### 7.1.2   *Slip rates in a megafault*

The discussion above has reviewed evidence for slip accumulation in the San Andreas fault. The type of accumulation may show considerable variation from place to place, but, in the long term, the fault tends to accumulate uniform slip over its whole length. Consequently, the contribution to the macroscopic fault movement by each seismic event, i.e. each partial slip, within the macroscopic fault, may be represented by averaging its amplitude over the total length of the megafault. Summing up the contributions from successive events over a long period, therefore, should give the cumulative slip of the whole fault.

Brune (1968) employed this idea and proposed a technique to calculate slip rate along a major fault zone on the basis of seismicity data. Let us assume the fault zone to be a simple surface resulting from opposing mass movements. Then, the slip amplitude, $U$, in an earthquake is given by:

$$U = M_o/\mu A, \tag{7.1}$$

(see (4.20) and (5.23)), where $\mu$ is the rigidity, $M_o$ is the source moment, and $A$ is the area of the fault slippage. Let $A_0$ denote the total surface area of the megafault to be studied, then we may translate $U$ for each fault of area $A$ into an effective homogeneous slip for the whole fault zone as follows:

$$\langle U \rangle = U\frac{A}{A_0} = \frac{M_o}{\mu A_0}. \tag{7.2}$$

All events are summed to get the long-term slip:

$$\Sigma\langle U \rangle = \frac{1}{\mu A_0}\Sigma M_o. \tag{7.3}$$

If $M_o$ is known for all the seismic events which occurred in a fault zone during a sufficiently long period of time, then (7.3) enables us to calculate the cumulative fault slip, and consequently, its time rate.

Brune introduced an empirical moment–magnitude relation, as shown in fig. 7.3, which he used to interpret (7.3) in terms of seismic magnitude. This is very useful because most earthquake statistics are based on the magnitude scale. Table 7.1 is an example of Brune's numerical work on seismicity in the Imperial Valley, southern California. The data for earthquake occurrence are taken from Allen *et al.* (1965) with $N$ denoting frequencies in each magnitude range during the period 1934–63 (values of $N$ for $M < 3$ are assumed). As $M_o$ is known from fig. 7.3, the total moment is estimated at $7.4 \times 10^{26}$ dyn cm by summing the products $NM_o$ over the whole range. To determine $A_0$, the depth of the zone is assumed to be 20 km. The length of the zone is 120 km, so the cross-

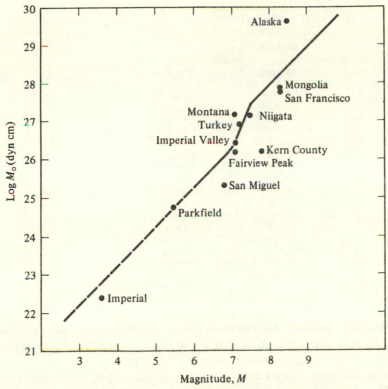

Fig. 7.3. The empirical relation of moment to magnitude. (After Brune, 1968.)

sectional area is $2.4 \times 10^{13}$ cm$^2$. Using (7.3) we obtain a total slip of 93 cm for the 29-year period, and a slip rate of 3.2 cm per year.

The geodetically observed rate is 4–5 cm per year in the same area, therefore, the calculated rate is roughly consistent with observation.

The example above proves the usefulness of (7.3) for an order-of-magnitude approximation. Brune (1968) applied this technique to several other major seismic zones and calculated their slip rates. Table 7.2 reproduces a part of his results. The rates for various parts of the San Andreas fault system are generally uniform with values of several centimetres per year. The minor San Andreas fault (i.e. in the narrower sense) and the Kern County fault have very low and high rates, respectively. These values seem to be due to inadequate sampling, however. The former area was almost aseismic in the period studied (1934–63) in spite of the great 1857 Fort Tejon earthquake. If this had been included, the calculated slip would have been several centimetres per year. Conversely, the high rate in the Kern County is mainly because of the occurrence of the Tehachapi earthquake and its aftershocks. The rate would have been low if the calculation had been made for the 29-year period preceding this

Table 7.1. *Earthquakes in the Imperial Valley,*
*1934–1963. (From Brune, 1968.)*

| Magnitude, $M$ | Moment, $M_o$ (dyn cm) | Number, $N$ | $NM_o$ (dyn cm) |
|---|---|---|---|
| 7.1 | $2.8 \times 10^{26}$ | 1 | $2.8 \times 10^{26}$ |
| $6\frac{3}{4}$ | $8.9 \times 10^{25}$ | 1 | $8.9 \times 10^{25}$ |
| $6\frac{1}{4}$ | $2.8 \times 10^{25}$ | 1 | $2.8 \times 10^{25}$ |
| $5\frac{3}{4}$ | $8.9 \times 10^{24}$ | 9 | $8.0 \times 10^{25}$ |
| $5\frac{1}{4}$ | $2.8 \times 10^{24}$ | 18 | $5.1 \times 10^{25}$ |
| $4\frac{3}{4}$ | $8.9 \times 10^{23}$ | 55 | $4.9 \times 10^{25}$ |
| $4\frac{1}{4}$ | $2.8 \times 10^{23}$ | 131 | $3.7 \times 10^{25}$ |
| $3\frac{3}{4}$ | $8.9 \times 10^{22}$ | 354 | $3.2 \times 10^{25}$ |
| $3\frac{1}{4}$ | $2.8 \times 10^{22}$ | 885 | $2.5 \times 10^{25}$ |
| $2\frac{3}{4}$ | $8.9 \times 10^{21}$ | 2 212 | $2.0 \times 10^{25}$ |
| $2\frac{1}{4}$ | $2.8 \times 10^{21}$ | 5 530 | $1.6 \times 10^{25}$ |
| $1\frac{3}{4}$ | $8.9 \times 10^{20}$ | 13 825 | $1.2 \times 10^{25}$ |
| $1\frac{1}{4}$ | $2.8 \times 10^{20}$ | 34 562 | $9.8 \times 10^{24}$ |
| $\frac{3}{4}$ | $8.9 \times 10^{19}$ | 86 405 | $7.7 \times 10^{24}$ |
| $\frac{1}{4}$ | $2.8 \times 10^{19}$ | 21 602 | $6.1 \times 10^{23}$ |
| Total | | | $7.4 \times 10^{26}$ |

earthquake. Brune remarked from these experiences that the calculated slip rates are strongly influenced by whether or not the time sample is long enough to represent the average seismicity.

The rate for the major San Andreas fault (6.6 cm per year) is calculated from a relatively long sample time and is probably acceptable in this

Table 7.2. *Calculated rates of slip for major fault zones. (From Brune, 1968.) The data in the*
*remarks column are from Le Pichon (1968).*

| Zone | Time Period | $L$ (km) | $D$ (km) | $\Sigma M_o$ (dyn cm) | Slip rate* (cm per year) | Remarks |
|---|---|---|---|---|---|---|
| South California | 1934–63 | | | | | |
| Imperial Valley | | 120 | 20 | $7.4 \times 10^{26}$ | 3.2 | |
| San Andreas fault (minor) | | 130 | 20 | $6.6 \times 10^{24}$ | 0.03 | |
| Kern County | | 134 | 20 | $4.4 \times 10^{27}$ | 17 | |
| Total area | | 754 | 20 | $8.3 \times 10^{27}$ | 5.8 | |
| San Jacinto fault | 1912–63 | 280 | 20 | $1.4 \times 10^{27}$ | 1.5 | |
| Southern California area | 1912–63 | 754 | 20 | $9.5 \times 10^{27}$ | 3.7 | |
| San Andreas fault (major) | 1800–67 | 1240 | 20 | $9.0 \times 10^{28}$ | 6.6 | |
| New Zealand | 1914–48 | 1350 | 20 | $2.2 \times 10^{28}$ | 7.2 | 1.7 cm |
| Turkey (Anatolian Fault) | 1939–67 | 1240 | 20 | $2.5 \times 10^{28}$ | 11 | per year |
| Island arcs | | | | | | |
| Tonga 0 to 60 km depth | 1920–54 | 1500 | 85 | $7.4 \times 10^{28}$ | 5.2 | 9.1 |
| Tonga 100 to 700 km depth | 1920–54 | 1500 | 800 | $5.7 \times 10^{28}$ | 0.23† | |
| Japan | 1905–55 | 800 | 60 | $1.2 \times 10^{29}$ | 15.7 | 8.8, 9.0 |
| Aleutians | 1905–67 | 3200 | 85 | $2.1 \times 10^{29}$ | 3.8 | 5.3–6.3 |

*Unless otherwise noted $\mu$ is assumed to be $3.3 \times 10^{11}$ dyn cm$^{-2}$.
†$\mu = 6 \times 10^{11}$ dyn cm$^{-2}$.

respect. The calculated rates are compared with calculated data by Le
Pichon (1968) for several seismic planes in the world (see the remarks
column in table 7.2). It is rather surprising that these two sets of data
show fairly good correlation, in spite of the different methods by which
data were obtained. This implies that the mechanism of slip accumulation in a megafault does not differ much from that supposed by Brune.

### 7.1.3 Earthquakes and plate tectonics

*Concepts of a plate.* Our contemporary view of global tectonics is
based on a hypothetical system of mobile *plates*, lying over the Earth's
surface and moving laterally continuously (e.g. Bird & Isacks, 1972; Le
Pichon, Francheteau & Bonnin, 1973; Lomnitz, 1974).

A tectonic plate is a relatively rigid layer (*lithosphere*) of the order of
100 km thick, which overlies the *asthenosphere* which is some hundreds
of kilometres thick. The asthenosphere has effectively no strength on a
long timescale and enables the plates to move in a relatively undeformed
manner. It corresponds more or less to the low-velocity layer, which is
characterized by strong attenuation (low $Q$-value) of seismic waves
passing through it (Fedotov, 1963; Utsu, 1966; Oliver & Isacks, 1967). If
the $Q$-value, which is the reciprocal of the specific attenuation factor,
gives a measure of the strength of the rock, the strong attenuation in the
low-velocity layer suggests 'weakness' of the layer.

Hypothetically, the low-velocity layer may be attributed to a critical
thermal state in the Earth such that the local melting point at depths
100–300 km from the surface is extremely close to the local temperature
(see Anderson, 1962). This presumably causes local decrease of strength,
and consequently, decreases in the seismic wave velocities in the layer. It
is appropriate that this layer is called the asthenosphere, from the Greek
root 'asthenes' which means 'weak'.

The properties of the lithosphere, from the greek root 'lithos' (stone),
contrast with those of the asthenosphere; i.e. its velocities, $Q$-value and
mechanical strength are high relative to those of the asthenosphere. The
lithosphere doesn't necessarily have a sharp bottom boundary, but may
gradually change in properties at a certain depth. It is relatively thin
about submarine ridges, and is likely to grow to a considerable thickness
as it moves toward the subduction zone (see for example Yoshii, 1973).
The geological features of a plate are very complex. However, we
usually sketch it, for simplicity, as a distinct and laterally rigid slab of
uniform thickness.

*Configurations of moving plates.* The version of plate tectonics
 . . y used by seismologists is concerned principally with the relative
  . . nt and interaction of adjoining plates. Fig. 7.4 shows various

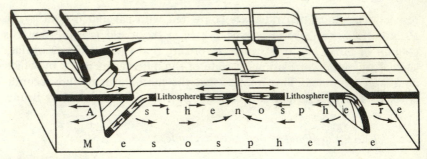

Fig. 7.4. Block diagram illustrating the configurations and roles of the lithosphere, asthenosphere, and the mesosphere in a version of the new global tectonics in which the lithosphere, a layer of strength, plays the key role. Arrows on the lithosphere indicate relative movements of adjoining blocks. Arrows in the asthenosphere represent possible compensating flow in response to downward movement of segments of lithosphere. One arc-to-arc transform fault appears at the left between oppositely facing zones of subduction, two ridge-to-ridge transform faults along ocean ridges appear at the centre, and a simple arc structure appears at the right. (From Isacks, Oliver & Sykes. 1968.)

configurations of moving plates over the asthenosphere and mesosphere. The major features are creation of plates at oceanic ridges (near the centre of the figure), and *subduction* at a trench (to the right), where the oceanic plate meets a continental plate and is forced beneath it. If we compare the figure with the south Pacific area, the ridges in the central part correspond roughly to the south and/or east Pacific rises, and the subduction zone (right), to western South America. The complex system to the left corresponds roughly to the New Hebrides, Fiji and Tonga area, where plate subduction from two different (almost opposite) directions occurs.

Another notable feature shown in the figure is that the crest of the ridge is discontinuous in a number of places. As the ridges grow and the plates move outward from the ridges (central part of the figure), a significant offset of strike-slip type can occur across a line segment which connects the adjoining crests of the ridge. This type of fault is called a *transform fault* after Wilson (1965). Fragmentary offsets of the ridge axes in the figure may appear to suggest right-lateral movements; but notice that spreading of plates from ridges assigns left-lateral slip across line segments connecting the ridge crests (see arrows in the figure).

Geological aspects of plate interaction are not as simple as in this idealized picture. The complex transform fault system about the Pacific coast of North America exemplifies this situation (§7.1.1). Geological complexities of this type will be considered in greater detail as our discussion proceeds into the details of the individual tectonic areas (see for example Sugimura & Uyeda, 1973). Also, the plate interaction scheme described above does not explain all earthquakes. The classificat' earthquake types by their plate tectonic environment is discussed briefly in this section.

*Evolution of a subduction zone.* An important aspect of the plate-tectonics theory is the geometrical kinematics of lithospheric plates over the Earth's surface. If the plate-tectonic theory is valid, then the possible modes of movements at the plate boundary, including ridges, trenches and transform faults, must be mutually restricted, so that they satisfy the condition of motion of a rigid plate. A convincing test of this was made by McKenzie & Parker (1967), taking the Pacific plate as an example. Fig. 7.5 shows the Mercator projection of the north Pacific area, about the pole fixed at 50°N, 85°W (see index map). Arrows in the map refer to the slip vectors from fault-plane solutions (see §3.2.3), and illustrate directions of the plate motion at the boundary sites, relative to the adjoining plate. Notice that the arrows are all parallel to horizontal co-latitude lines, which is a necessary condition for rigid plate rotation about the pole given above.

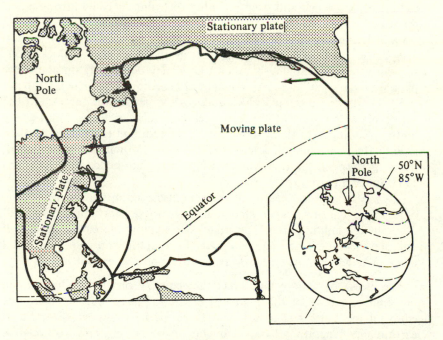

Fig. 7.5. A Mercator projection of the Pacific with a pole at 50°N 85°W. The arrows show the direction of motion of the Pacific plate relative to that containing North America and Kamchatka. If both plates are rigid all slip vectors must be parallel with each other and with the upper and lower boundaries of the figure. Possible boundaries of other plates are sketched. (After McKenzie & Parker, 1967.)

The mechanical behaviour of plates undergoing subduction can be studied seismologically, on the basis of the topographical distribution of earthquakes in a seismic plane, and the slip type determined from source-mechanism studies. Kanamori (1977) performed a comparative study of the seismological features of subduction zones around the Pacific plate. Fig. 7.6 shows cross-sectional views of plate subduction in several representative areas. The views are arranged in an order that, we speculate, represents a coherent process of evolution at a subduction zone. It is supposed that a gravitational instability, caused by progressive cooling of the oceanic plate, drives the evolution as follows:

(a) The oceanic plate meets the continental block. Cooling of the oceanic plate makes it dense, even in comparison with the asthenosphere. But the plate could remain horizontal for a certain while, due to mechanical support by the continental block. Sooner or later, the gravitational instability exceeds a critical value, and the heavy plate starts sinking, thrusting repeatedly against the bottom slope of the continental wedge. Earthquakes, such as the 1964 Alaskan earthquake, caused by low-angle and deep-reaching thrusting, may be attributed to this stage.

(b) Frictional heating may supply energy for volcanic activity. It may also cause subcrustal partial melting in the continental block; consequently, seismic slipping between plates may occur only on a shallow plate boundary (e.g. the 1968 Tokachi-oki earthquake, northeast Japan).

(c) The partially molten block reduces its mechanical support to the heavy plate. As a result the plate tends to accelerate its down-going motion, leading to further subduction. If the resultant tension in the plate exceeds the plate strength, a tensile fracture, i.e. normal-type faulting, will occur, perhaps in the trench area, where the plate bends most (e.g. the 1933 Sanriku earthquake, Japan).

(d–f) After tensile fracture, of the sort discussed in (c), the lower part of the plate will sink deeper. Deep and intermediate earthquakes will still occur in the falling part of the plate, but shallow big earthquakes will not, since the oceanic plate has little interaction with the continental block. The activity in the Izu-Mariana arcs may be caused in this way.

The theory of plate-tectonics has allowed great progress to be made in our understanding of the many diverse disciplines that make up the study of the Earth's tectonics. The basic question about the origin of earthquake generating forces can be answered in terms of relative movement and interaction between the adjoining plates. If we are to go into this problem in more detail, however, we must face the unanswered question of what mechanisms cause the plates to move. Mantle convection was considered the most plausible cause in the early years of plate-tectonic theory. Recently, however, it has been replaced by models

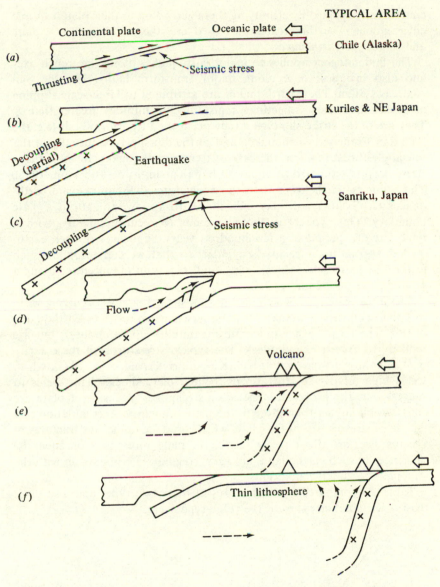

Fig. 7.6. Model of plate subduction evolution. The stages develop from top to bottom. (After Kanamori, 1977.)

based on gravitational instability. Further investigation of the behaviour of plates is needed for this critical problem to be resolved.

*Inter- and intra-plate earthquakes*. The theory of plate-tectonics suggests that the essential characteristics of shallow earthquakes may be conveniently typified by classifying them according to their plate-tectonic environment. The three categories used are, the *inter-plate, intra-plate* and *ridge-type earthquakes.*

The first category applies to major earthquakes along the trench axes and also includes those along major transform faults, like the San Andreas system. These earthquakes are attributed to large-scale slipping at a plate boundary, sometimes registering magnitudes greater than 8. They are of the strike-slip type if they occur on a transform fault (e.g. the 1906 San Francisco earthquake), and of the dip-slip type (usually of the low-angle thrust type) in the case of trench earthquakes. In both cases, stress-drop is estimated at about 30 bar (Kanamori & Anderson, 1975). This value is much lower than those determined for intra-plate earthquakes (see below), and may indicate relatively low strength at a plate boundary. This type of earthquake tends to recur frequently, with a typical return period of a few hundred years. Sequences of such earthquakes appear to be relatively regular, so that we can recognize such features as periodicity, seismic gaps, and migration of epicentres along a seismic zone (§7.2.2 and §7.2.3).

The intra-plate type, which includes most earthquakes occurring within a plate, contrasts with the inter-plate type in many respects. Earthquakes of this type may register moderate magnitudes up to about 7, but are unlikely to exceed magnitude 8. The typical stress-drop in these earthquakes is approximately 100 bar (Kanamori & Anderson, 1975), which seems high relative to stress-drops in inter-plate shocks. This seems to suggest that the frictional stresses, or strength of rocks, in a plate are considerably higher than those in the plate boundary. Accumulation and release of stresses within the plate, which produce numerous fractures of various size, are not directly related to plate motions. Consequently, sequences of such earthquakes are very irregular, in comparison with the sequences for inter-plate events.

Table 7.3 compares typical features of these two types and also gives the known characteristics of the ridge-type event.

## 7.2   RECURRENCE OF EARTHQUAKES

### 7.2.1   *Strain accumulation*

*Strain and stress across a fault.* The discussion in §7.1.1 has shown that the long-term land offset across the San Andreas fault can be interpreted by the accumulation of repeated coseismic slips. Fig. 7.7 shows an idealized model of accumulation in which slipping is repeated with uniform amplitude at regular time intervals. Fig. 7.7*b*, showing offset against time, corresponds to the stress changes shown in fig. 7.7*a*.

Table 7.3. *Classification of shallow earthquakes according to their plate-tectonic environments.*

|  | Inter-plate | Intra-plate | Ridge-type |
|---|---|---|---|
| Place of occurrence | subduction zone, major transform fault | plate (body) | oceanic ridge |
| Possible magnitude ($M$) | $\approx 8.5$ | $\approx 7.5$ | small, in general |
| Stress-drop | $\approx 30$ bar | $\approx 10^2$ bar | ? |
| Recurrence | $\approx 10^2$ years | $10^3 - 10^4$ years | ? |
| Remarks | regularity (recognizable periodicity, seismic gap) | complex |  |
| Examples | San Francisco (1906) Kanto (1923) Chile (1960) Alaska (1964) | Tango (1927) Niigata (1964) |  |

The range of stress change is specified by the two constants, $\sigma_s$, static frictional stress, and $\sigma_k$, dynamic (or, kinematic) frictional stress (these symbols differ from those used in §6.2.3, as these quantities are considered in a broader sense here).

A more realistic diagram may be drawn by introducing some irregular features, as shown in fig. 7.8. Now, the basic question is which of the frictional stresses, $\sigma_s$ or $\sigma_k$, is more intimately associated with the strain

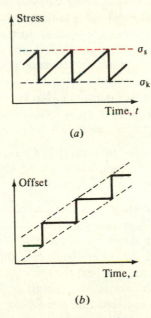

(a)

(b)

Fig. 7.7. (a) Accumulation of stress; and (b) offset at a fault. $\sigma_s$ and $\sigma_k$ denote, respectively, the static and dynamic (kimenatic) frictional stresses in a fault surface.

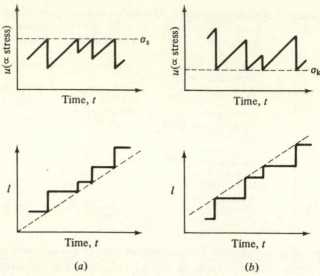

Fig. 7.8. Two extreme cases of faulting in terms of the stress condition $u$ and $l$ represent displacements defined in fig. 7.9. (After Shimazaki, 1977.)

processes. Let us study this problem in more detail, as it might prove to have useful bearings on prediction studies (see §7.2.2 and §8.2).

A conveyor-belt with a mass being carried on it (fig. 7.9) is used to model two adjoining land blocks at a fault, which accumulate occasional slips under the frictional stresses, $\sigma_s$ and $\sigma_k$. In the figure, $l$ and $u$ denote displacement of the mass relative to the belt (representing slip across the plate boundary) and to a stationary reference point (representing deformation of the continental block) respectively (Shimazaki, 1977). The mass is linked to the reference point via a flat spring (elasticity of the land block), and therefore $u$ is proportional to the crustal stress in the block.

Let us assume two extreme cases;

(a) slip occurs only when the crustal stress exceeds $\sigma_s$;
(b) slip may occur at a moderate unfixed stress level, but, once it happens, it will not cease until the crustal stress decreases to $\sigma_k$.

Fig. 7.8a and b represent these two cases respectively.

Let us assume a uniform rate of strain accumulation. If case (a) is assumed then the slip amplitude in an event will specify the time interval to the next event. If, on the other hand, case (b) is assumed, the lapse time since the last event specifies the potential slip amplitude at a particular time. In other words, the timing of the next event can be predicted relative to the previous one (see §7.2.2), if the slip amplitude of the future event is specified (Shimazaki, 1977).

Fig. 7.10 shows the cumulative moment and seismic slip at the

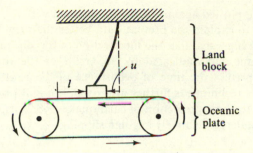

Fig. 7.9. A model of land block movement in a fault. (After Shimazaki, 1977.)

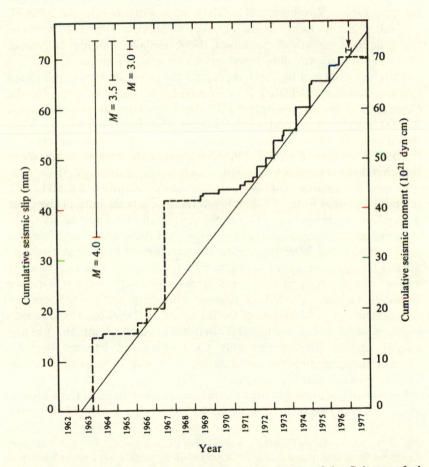

Fig. 7.10. Cumulative moment and seismic slip in a zone of the Calaveras fault (1962–77). The arrow at the top right indicates the anticipated time of intersection of the slip-rate line and the projected seismic slip assuming an average rate of small ($M \le 2.5$) earthquake activity in the interim. (From Bufe, Harsh & Burford, 1977.)

Calaveras fault, plotted against time. Comparison with fig. 7.8 shows that case (*a*) seems to explain the present data better than (*b*). Bufe, Harsh & Burford (1977) suggested that the intersection of the slip-rate line and the projected seismic slip (indicated by an arrow at the top right of the figure) might predict the time of occurrence of the next earthquake in this area. If this technique is further improved, it might provide us with a useful tool for tectonic prediction of earthquakes (§8.2), especially in fault zones which undergo relatively regular slipping.

*Strain and stress fields.* The set of charts in fig. 7.11 gives a comprehensive picture of strain and stress fields in the central part of Honshu, Japan (Kasahara, 1971). They are constructed on the basis of: (*a*) fault-plane solutions of earthquakes (Ichikawa, 1966); (*b*) triangulation data (Kasahara & Sugimura, 1964; see also Harada & Kassai, 1971); and (*c*) tectonic data from various sources (Matsuda, 1967).

The chart shown in fig. 7.11*b* represents the principal axes of minimum horizontal strain calculated from triangulation data collected by the Geographical Survey Institute (GSI). Readings were taken over several decades with short intervals between them. The chart shown in fig. 7.11*a* shows pressure axis directions (§3.2.3) for very shallow earthquakes for the period between 1926 and 1965. Notice that the axes shown in these two charts coincide at most places.

It must be remarked that the pressure axis directions in fig. 7.11*a* and the principal axes in fig. 7.11*b* indicate the mechanical state of the crust from different viewpoints. The former represents the direction of force, changing at the moment, and at the locality, of the seismic fracture in the crust, whereas the latter represents the direction of predominant deformation (contraction) of the medium, either seismic or aseismic, which is averaged over long periods of time and over a wide area. It is rather surprising that such excellent agreement is found in spite of the complex crustal structures and in spite of the disturbance of the strain field caused by the several major earthquakes that took place during the observational interval. The unexpectedly good agreement between the two patterns indicates that the mechanisms of crustal movements may be simpler than is generally thought.

Further information on crustal stresses, which determine the tectonics in an area, may be inferred from geological evidence such as the strike

Fig. 7.11. Maps of the central part of Japan comparing: (*a*) pressure axis directions from fault-plane solutions of very shallow earthquakes (from Ichikawa, 1966); (*b*) directions of greatest relative shortening inferred from the triangulation data by the Geographical Survey Institute (from Kasahara & Sugimura, 1964); and (*c*) the stress field (lines represent directions of compressional stress), inferred from various tectonic events such as active faults, active foldings, arrangement of fissures in volcanic bodies, etc. (From Matsuda, Okada & Huzita, 1976.)

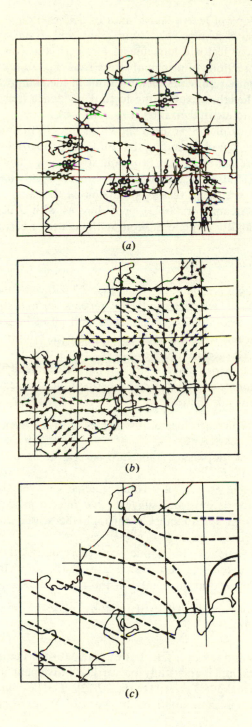

(a)

(b)

(c)

and slip directions of active faults, axes of active foldings, alignment of craters, direction of fissures in volcanic areas, and so on. Matsuda (1967) studied the recent tectonic movements in Japan and compiled regional trends of the directions of principal stress axes (fig. 7.11c). The family of curves in the figure represents the directions of the compressive axes, and are drawn so that they explain the tectonic evidence (seismological and geodetic data are also taken into consideration).

The consistency of the patterns shown in figs. 7.11a, b and c strongly supports our basic understanding that all these properties in the area are caused by a persistent (cf. §7.2.2) tectonic stress which tends to compress the Japanese Islands in a northwesterly direction. This trend of stress is also consistent with slip vectors of major earthquakes off the Pacific coast of Japan (Kanamori, 1973), providing us with a persuasive argument for the validity of plate motion as the cause of earthquake events.

### 7.2.2   Seismic cycle

*Evidence for seismic recurrence.* Fig. 7.7 shows a fundamental type of fault movement, which periodically repeats seismic slips of uniform amplitude and accumulates a significant land offset in a long period of time. Historical records from several seismic areas confirm this idea. For example, fig. 7.12 shows a sequence of major earthquakes off the Pacific coast of Honshu, Japan since A.D. 684. The circles linked with thick lines represent the approximate location and size of the sources. Several points are evident from this figure:

(a) seismic recurrence, or seismic cycles with an average interval of about 100–200 years;

(b) division of the seismic zone into several areas with fixed boundaries, A, B, C, D, E;

(c) recognizable epochs of seismic quiescence over the whole zone, separating active periods; this suggests that once a partial break occurs, breaking tends to spread quickly through the whole zone (notice the predominant trend from east to west);

(d) similar frequencies of seismic occurrence across the whole seismic zone, suggesting uniform slip over a megafault. A northwestward slip direction is supposed at all points on the zone (see Kanamori, 1973).

All these points are well explained by plate motion in the plate-tectonic theory.

*Aspects of a seismic cycle.* Let us take Muroto Peninsula, Shikoku, Japan as an example, and study its vertical movement during the 1946 Nankai earthquake ($M = 8.1$; see fig. 7.12). The levelling data in fig. 7.13 show the coseismic uplift of the Muroto Promontory to be about

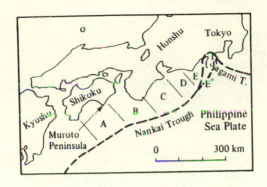

| Region | A | B | C | D | E | Sagami Trough |
|---|---|---|---|---|---|---|
| A.D. 684 | o—— | o—— | ? —— | ? —— | ? | 818, $M = 7.9$ |
| 887 | o—— | o—— | ? —— | ? —— | ? | |
| 922 | | | ● | | | |
| 1096 | | | o—— | o—— | ? | |
| 1099 | o——o | | | | | |
| 1360 | | ?—— | ● | | | |
| 1361 | o——o | | | | | |
| 1403 | | | ? | | | |
| 1408 | | | ● —— | ? | | |
| 1498 | | | o—— | o—— | E'? | |
| 1520 | | | ● | | | |
| 1605 | o—— | o—— | ? —— | ? —— | E'? | 1703, $M = 8.2$ |
| 1707 | o—— | o—— | o—— | o—— | o | |
| 1854 | | | o—— | o—— | o | |
| 1854 | o——o | | | | | 1923, $M = 7.9$ |
| 1944 | | | o—— | o | | |
| 1946 | o——o | | | | | |

*Year of event* (vertical axis label, left side)

o  $M \approx 7$

● $M \approx 8$

Fig. 7.12. Recurrence of great earthquakes along the Nankai Trough. The circles and line segments in the lower diagram denote the ranges of the seismic source regions. The trough branches at its northern end and the western and eastern branches are denoted E and E' respectively. (After Utsu, 1976a, b; Aoki, 1977.)

120 cm relative to a stable station at the northern end of the peninsula. Also notable is the postseismic subsidence, which decelerates gradually to a rate of about −7.5 mm per year (see fig. 7.15). This rate of movement is consistent with the rate of subsidence that had been continuing for many years prior to the earthquake. The interseismic movement is well represented by a linear trend of subsidence at the rate stated above. The land movement in a seismic cycle may be interpreted in terms of four basic phases, i.e. the *interseismic, preseismic, coseismic* and *postseismic* phases. The preseismic phase will be discussed in more detail in the following chapter.

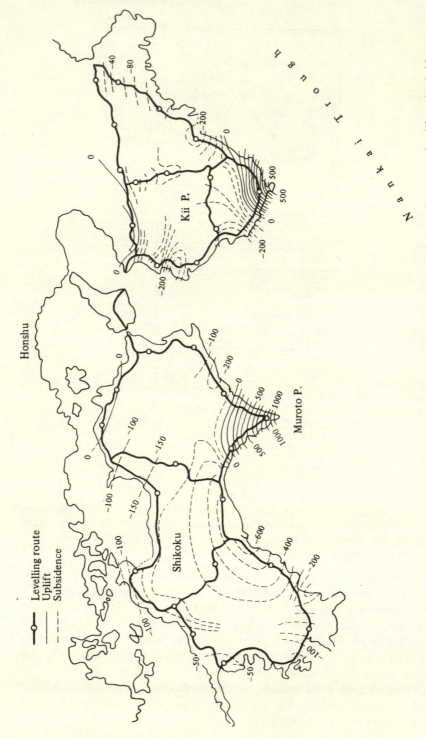

Fig. 7.13. Coseismic vertical land movement in the 1944 Tonankai and 1946 Nankai earthquakes. Contours of uplift and subsidence are labelled in mm. (From Miyabe, 1955.)

The uplift or subsidence of the Muroto Promontory in fact represents the motion of the whole peninsula, tilting like a rigid board about an east–west axis at its northern end. We can envisage a model of the subducting oceanic plate dragging a land block down with it during the interseismic period until a seismic rebound occurs, to explain the observations. A schematic view of land movement, as inferred from geodetic and geomorphic evidence (see below), is shown in fig. 7.14. Note that the cyclic repetition of a seismic uplift and interseismic subsidence produces, on a very long timescale, a secular trend of uplift (i.e. tilling away from the trench axis), which cannot be explained by the dragging model. Thickening of accretionary prisms at the continental slope (Nakamura, 1977) might provide an explanation, although little is known about this process.

Fig. 7.14 interprets the accumulation rate, $s$ of displacement (or tilt) in terms of three tectonic parameters: coseismic displacement, $U$; recurrence time, $T$; and *recovery rate*, $r$, which is the ratio of the integrated post- and interseismic subsidence in one cycle to $U$. These parameters are related (Matsuda, 1976) by

$$T = (1 - r)U/S. \tag{7.4}$$

If $r = 0$, the curve in fig. 7.14 resembles that in fig. 7.7b but this resemblance is deceptive, since the ordinate in fig. 7.14 represents the movement relative to sea level, not to the land block on the other side of a fault.

Geomorphic studies of sea-terraces along the coast of the Muroto Peninsula have shown that the land has accumulated a northward tilt

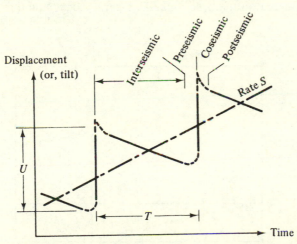

Fig. 7.14. A schematic representation of the vertical land movement (or, tilt) at Muroto Promontory showing the seismic cycles. The symbols are explained in the text.

(away from the trench) of about $1.1 \times 10^{-2}$ rad in the last $1.7 \times 10^5$ years (resulting in an uplift of the southern tip of about 350 m).

This suggests lone-term rates of $0.7 \times 10^{-7}$ rad per year in tilt and 2 mm per year in uplift (Yoshikawa, Kaizuka & Ota, 1964). If we assume that $U = 120$ cm as in the 1946 event, we obtain from (7.4) that $T/(1-r) = 600$ years. Accordingly, we take as a possible set of parameters, $T = 150\text{--}200$ years and $r = \frac{2}{3} - \frac{3}{4}$, which seems reasonable considering the evidence given in fig. 7.15.

*Chronology of recent crustal movements.* As a result of a geomorphic study of Awashima, a small island located close to the epicentre of the Niigata earthquake, Honshu, Japan (1969, $M = 7.5$; see fig. 8.7), an important chronological feature of the tectonic movements in this area was discovered. Table 7.4 compares geomorphic data of tilted sea-terraces of different ages with the coseismic movement in the 1975 event (§5.3.2 and §8.3.2), in which the island experienced an uplift of about one metre and tilt of about 1 minute of arc toward WNW (strike: N29°E; Nakamura, Kasahara & Matsuda, 1964). Note that the island has accumulated tilt of similar trend (striking NEN) for many thousands of years at a uniform rate, $S = 0.1''$ per year. This suggests that $T = 600$ and 300 years if $r = 0$ and $\frac{1}{2}$, respectively (from (7.4)).

The base rocks in this area provide the oldest evidence for tilting and are tilted by 10–20°. This suggests a duration time for the trend of about one million years (more precisely, $3.6\text{--}7.2 \times 10^5$ years. The age of this rock (dolomite) is thought to be $2\text{--}3 \times 10^7$ years (Miocene). Constant tilting at the rate given above would' produce a much larger overall tilt than that observed on the base rock. This paradoxical situation leads us

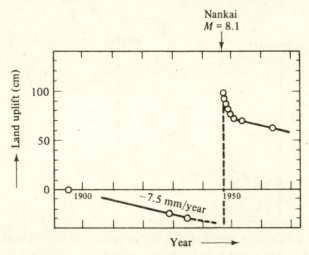

Fig. 7.15. Vertical land movement at Muroto Promontory observed by levelling. (From Okada & Nagata, 1953.)

Table 7.4. *Summary of geomorphic data from Awashima Island, approximately 10 km north-east of the epicentre of the Niigata earthquake (1964). (From Nakamura, Kasahara & Matsuda, 1964.)*

| Tilted layer | Strike | Tilting | Age | Rate | |
|---|---|---|---|---|---|
| Base rock (dolomite) | N30°E | 10–20° | Miocene | | |
| Sea terrace A | N25°E | 2° or more | 80 000–90 000 years | 0.07" per year | approx. 0.1" |
| Sea terrace B | N30°E | 15–30′ | c. 10 000 years | 0.09–0.18" per year | per year |
| Coseismic (1964) | N29°E | 56" | | | |

to suppose that the present mode of tilting has been active only in the last million years, and was preceded by a period of relative inactivity since the formation of the base rock. Data from several other tectonic areas in Japan seem to support this speculation. Fig. 7.16 plots the cumulative movement (e.g. fault offset) against age for several tectonic areas and the suggested history can be recognized.

Note that the Japanese data in fig. 7.16 look significantly different from those of the San Andreas fault in terms of duration time. This implies that tectonics in various regions may have different timescales and that the word 'recent' can only be understood in terms of how far in the past the patterns can be traced. The systematic and comprehensive study of regional tectonics with especial emphasis given to crustal

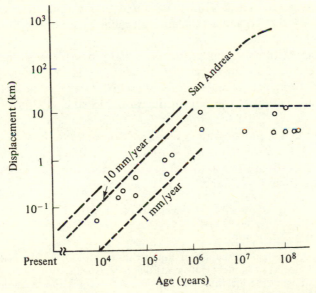

Fig. 7.16. Time–displacement diagram of major strike-slip faults in Japan, compared with movement at the San Andreas fault. (From Matsuda, 1977.)

movements and seismic activity is usually called the study of *recent crustal movement* where the word 'recent' is to be understood in the sense explained above.

*Active faults.* In the same way that volcanoes can be classified as active and extinct, faults in a geological field may be classified according to whether they are potentially active or not. The term *active fault* is used most commonly to denote a fault whose movement in Quaternary time has been recognized. This definition is based on the understanding (§7.1 and §7.2) that Quaternary tectonics are still continuing at the present time, in a macroscopic sense.

The degree of activity of a fault may sometimes be assessed by geological means (e.g., the spatial relation of the fault's offset to geological formations of known ages), and sometimes by geomorphic means (e.g., offset of terraces and stream courses, as well as the ages of fault scarfs). Historical or instrumental records can also be used when available.

Fig 7.17 shows maps of active faults compiled by these methods for: (*a*) Japan; and (*b*) New Zealand. In the northeastern part of Honshu, Japan, reverse faults striking north–south predominate. In the central part of Japan, numerous strike-slip and reverse faults predominate. Faults are thinly distributed over the southwestern part of Honshu, Shikoku and Kyushu, although the Median Tectonic Line has been recognized recently as the longest and dominant inland fault in this region. Information about off-coast faults is extremely poor, but the series of major faults along the Kurile–Kamchatka and Japan Trenches, as well as those along the Sagami and Nankai Troughs (see figs. 7.12 and 7.18), should be mentioned.

Assessment of active faults can potentially provide basic information about seismic risk. In order to answer various questions posed by earthquake prediction and hazard reduction work, we need to include the tectonic parameters in our formulation. Matsuda (1975), for example, classified active faults according to their long-term slip rate, $S$, as shown in table 7.5. Appendix 2 shows empirical formulae derived by a number of workers, from which we can see that $U$ in (7.4) depends on magnitude $M$, and accordingly, on fault length $L$. This allows us to express the recurrence time $T$ as a function of $M$ and $S$, as

$$\log T = 0.6M - (\log S + 4.0), \tag{7.5}$$

---

Fig. 7.17. Active fault maps: (*a*) of Japan, showing the position of the Median Tectonic Line (MTL) (after Matsuda, Okada & Huzita, 1976); and (*b*) of New Zealand, showing the location and occurrence time of major earthquakes (after Lensen, 1965).

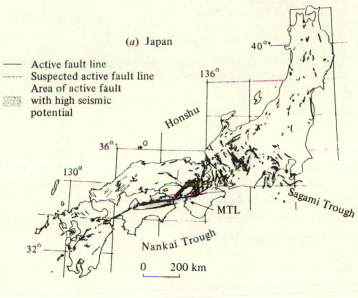

(*a*) Japan

— Active fault line
----- Suspected active fault line
Area of active fault
with high seismic
potential

40°

136°

Honshu

36°

130°

32°

Sagami Trough

MTL

Nankai Trough

0     200 km

(*b*)  New Zealand

1914

1932

1931

1897

North I

1904
1934
1942

1929

W

South I

1848  1855

1929

H

1901
1888

Alpine Fault

1960

M

— Active fault line
• Epicentre of
major earthquake

0     200 km

Table 7.5. *Classification of active faults based on long-term slip rates. (From Matsuda, 1975.)*

| Class | S (mm per year) | Examples of faults |
|-------|------------------|--------------------|
| AA | $10 \leqq S < 100$ | Nankai Thrust (off Shikoku) |
| A | $1 \leqq S < 10$ | Median Tectonic Line, Atera |
| B | $0.1 \leqq S < 1$ | |
| C | $0.01 \leqq S < 0.1$ | |

where $T$ and $S$ are measured in years and m per year respectively (Matsuda, 1977). The seismic recurrence time estimated by empirical formulae such as (7.5) is subject to considerable error, yet it provides information on which to base more precise studies (§8.1.2).

### 7.2.3   *Seismic gap*

*Sequence of inter-plate earthquakes.* Triggering of an earthquake is generally thought to be a stochastic process, so that regularities in seismic activity, if they exist, may be recognized only on a statistical basis (e.g. magnitude–frequency relation, laws of aftershock sequence, etc.). Yet, major earthquakes may sometimes occur in a relatively systematic manner, as if their occurrence is deterministically controlled by mutual interaction. Examples are the systematic migration of earthquakes along a major seismic zone (§7.3.2), and also, the empirical laws of a seismic gap, which will be discussed in the following. There is still great uncertainty about the mechanisms that cause these remarkable characteristics, but stationary stressing of a seismic zone by plate motion and finiteness of the seismic field relative to the source size are likely candidates. The latter implies that occurrence of a major earthquake will significantly disturb the space around it, as its source dimension is comparable with the size of a seismic zone.

Several workers, for example Fedotov (1965), noticed that aftershock regions in major trench earthquakes spread along the axis of the trench in such a manner that the whole zone is uniformly covered in one seismic sequence. Fig. 7.18 demonstrates the arrangement of aftershock regions along the Kurile–Kamchatka Trench (see also fig. 7.12 for the earthquakes off Honshu, and fig. 7.21 for the Anatolia earthquake series).

Note in fig. 7.18 that aftershock regions in the seismic sequences between 1952 and 1970 covered almost the whole length of the seismic zone, with only one vacant area (number 8) off the southeastern part of Hokkaido. That is to say, region 8 remained a *seismic gap* (or, *seismicity gap*). An aftershock region is generally thought to well approximate the region of main fracture (§2.3.2). Therefore, in conjunction with evidence

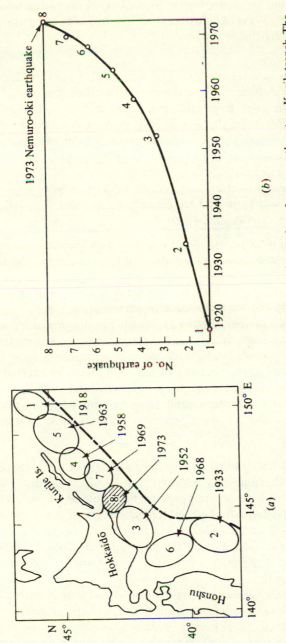

1973 Nemuro-oki earthquake

(b)

No. of earthquake

150° E

1918
1963
1958
1969
1973
1952
1968
1933

Kunile Is.

Hokkaidō

Honshu

145°

(a)

140°

N

45°

40°

Fig. 7.18. (a) Locations of focal regions of large shallow earthquakes along the northern Japan–southeastern Kurile trench. The number assigned to each earthquake shows the order in time of its occurrence. (b) The earthquake sequence is shown as a function of time. (From Mogi, 1977.)

of accumulation of intense strain in the southeastern part of Hokkaido, it was very reasonable for seismologists to predict an imminent break. In fact, in 1973, the Nemuro-oki earthquake ($M = 7.4–7.7$) occurred exactly in this area, fulfilling the prediction.

*Recognition of a seismic gap.* The successful prediction discussed above proved the usefulness of the seismic gap method. Since that time, many other seismic gaps, large and small, have been found in different areas. If the application of the seismic gap method is widened without proper care, unnecessary confusion might result. Closer examination of the Japanese success is useful to ensure the proper application of this method to future cases.

Prior to the Nemuro-oki earthquake, an inter-plate seismic gap was recognized which was large and tectonically meaningful. The following points support the validity of the gap:

(*a*) the topographical situation of the gap, which seemed to occupy a reasonable area in the seismic plane, dipping landward from the trench axis;

(*b*) the size of the gap, which was sufficiently large, and was compatible with a potential seismic source of magnitude 7.5–8;

(*c*) the recurrence of seismic activity, which was confirmed by historical records; and the lapse time, which was sufficiently long since the last event of 1894;

(*d*) the presence of seismic activity of the whole zone, as inferred from earthquake occurrence in adjacent areas, since the early 1950s;

(*e*) the intense strain accumulation (observed inland) over previous two decades, in a sense consistent with the regional stress (northwest–southeast compression).

The mechanical background of the empirical law of a seismic gap is presumably the behaviour of a plate tending to slip uniformly over its boundary surface with an adjoining plate. For the case of uniform slip, with little internal deformation of a plate, an area at the interface cannot remain unmoved for long. It is easy to imagine, from this viewpoint, that the accuracy of the law depends essentially on the mechanical differences between a plate and the interface medium, indicated by differences in parameters such as the rigidity, strength, etc.

Anomalous seismic quiescence may be found to occur persistently, or temporarily (e.g. ceasing of seismic activity at the microearthquake level, preceding a moderate shock), even within intra-plate seismic regions. When we call them seismic gaps, however, special care must be taken so that the concepts of an inter-plate seismic gap are not used inappropriately.

## 7.3   ANELASTIC ASPECTS OF CRUSTAL MOVEMENTS

### 7.3.1   *Aseismic movements of the crust*

Classical work on post-glacial uplift in Fennoscandia yielded the first successful measurement of mobility within the Earth (cf. Heiskanen & Vening Meinesz, 1958). Since that time, extensive work has been conducted on anelastic aspects of land movements, in which mobility or plasticity of the Earth's medium play an essential role. The most impressive demonstration of such mobility is the evidence accumulated for the explanation of continental drift in terms of plate movement over the mobile asthenosphere (§7.1.3). In this section, however, we shall study local movements related to aseismic fault slips.

Fault creep near Hollister, central California (see fig. 7.1), is a famous example of this kind of phenomenon. Fig. 7.19 shows accumulative land offset across the fault, determined from deformation of linear man-made features of various ages. Note that the land has accumulated offset (right-lateral) over the last 60 years at an almost constant rate of approximately 2 cm per year. This figure includes a small number of data representing seismic movement in local earthquakes, but most of the movement is not associated with appreciable earthquakes. It is evident that the segment of the Calaveras fault undergoes almost aseismic creep for a considerable length about Hollister. The mechanism for this is not fully known, but it

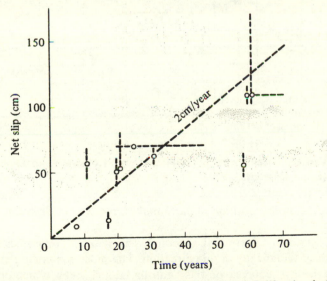

Fig. 7.19. Approximate rate of slip at Hollister, central California, derived from studies of linear man-made features of various ages. Horizontal and vertical bars represent errors in the age and slip amplitude respectively. (After Brown & Wallace, 1968.)

may reflect a special rockmechanical condition of this segment which lies between two segments where repeated seismic slip occurs (e.g. the great earthquakes of 1857 and 1906). The segment has continued to creep at an almost constant rate for an indefinite period (Brown & Wallace, 1968).

Postseismic recovery of land movement, which is often observed in major earthquakes off the Pacific coast of Japan, may also yield evidence supporting the hypothesis of anelastic fault movement (see fig. 7.15 for the Nankai earthquake, 1946). Fig. 7.20 shows the vertical land movement at the southeastern tip of Hokkaido (station: Hanasaki) over a period of about eighty years between the two Nemuro-oki earthquakes in 1894 and 1973 (region 8 in fig. 7.18). The data indicate a general trend of land subsidence at a very high rate ($-8$ to $-9$ mm per year) over the last seventy years, but a significant trend of uplift for the first ten years. This anomalous effect seems to suggest delayed postseismic slip on a deeper part of the main thrust fault.

Anelastic faulting behaviour may also be seen in tsunami earthquakes (§5.4.4) and preseismic land movements (§8.3.2). Our knowledge about these problems is generally very poor, in spite of their importance in both basic and applied seismology.

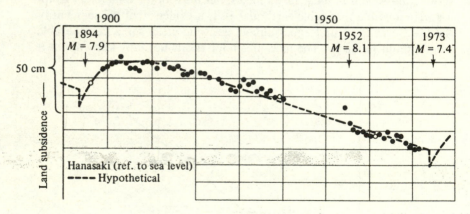

Fig. 7.20. Vertical movement at Hanasaki, southeast Hokkaido, Japan, relative to sea level. (From Kasahara, 1975.)

### 7.3.2. *Slow migration of earthquakes and crustal deformations*

*Observations.* Earthquake foci in a seismic zone sometimes appear to migrate systematically in one direction. The most convincing evidence for this is the last sequence on the Anatolia fault, Turkey, which occurred between 1939 and 1967. Fig. 7.21 shows the epicentres of these major earthquakes, and it is easy to see the trend of migration from the eastern to the western extremity of the fault (Richter, 1958; Mogi, 1968a). If we

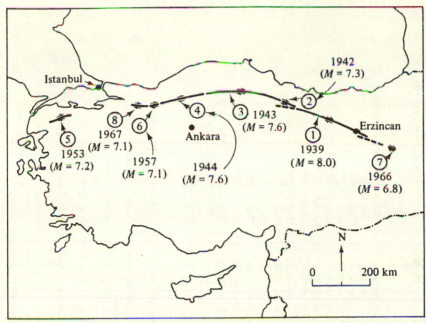

Fig. 7.21. Fault displacements along the North Anatolia fault associated with major historic earthquakes since 1939. (From Allen, 1968.)

assume a fault of moderate length about the epicentre (i.e. 50 to 200 km for magnitude of 7.5 to 8), then the sum of these is compatible with the whole fault extension (about 1500 km). Therefore, we may postulate a pattern of progressive fracturing in the Anatolia fault, which took almost 20 years to complete. The mechanism that controls such a slow propagation is unknown, but let us take the apparent velocity of epicentre migration (about 80 km per year in this case) and call it the *migration velocity* for future reference.

Following the discovery of such a pattern in Turkey, further examples of a similar kind have been recognized in several other areas by Mogi (1968a,b, 1969, 1973), Kelleher (1972) and Whitcomb *et al.* (1973), although in some of the examples the pattern is not as obvious as in the Turkish case. Table 7.6 (Kasahara, 1979) gives details of earthquake migrations so far reported. Note that the migration velocity appears generally to be in the range of 10–100 km per year except for a few examples of local aftershocks which have much higher migration velocities. Such good agreement in migration velocities may not be coincidental. Thus, several theoretical models have been proposed to account for the mechanisms which control the slow migration.

Migration of crustal deformation, recently discovered by tiltmeter and strainmeter observations, provides us with a new aspect of the phenom-

Table 7.6. *Velocity of migration.* (*From Kasahara, 1979.*)

| Type of event | Locality | Period | Velocity | Reference |
|---|---|---|---|---|
| Major (shallow) earthquakes | Anatolia (Turkey) | 1939–57 | 80 km per year | Richter (1958), Mogi (1968a), Savage (1971) |
| | Philippines | 1930–60 | 50 km per year | Mogi (1973) |
| | Kita-Izu (Japan) | 1930–62 | 12 km per year | Mogi (1969) |
| | Sanriku (Japan) | 1926–65 | 150 km per year | Mogi (1968a) |
| | West coast of US | 1900–70 | 60 km per year | Savage (1971) |
| | Chile | 1880–1960 | 10 km per year*** | Kelleher (1972) |
| Deep earthquakes | Mariana arc | 1930–65 | 50 km per year | Mogi (1973) |
| | Tonga arc | 1900–69 | 45 km per year | Mogi (1973) |
| Aftershock sequence | Aleutian | 9.3.1957 | 400 km per hour | Mogi (1968b) |
| | Alaska | 27.3.1964 | 60 km per hour | Mogi (1968b) |
| | San Fernando | 1971 | 4–15 km per day | Whitcomb et al. (1973) |
| Fault creep | Central California | 1971 | 1–10 km per day | King, Nason & Tocher (1973); Nason & Weertman (1973) |
| Major earthquakes seismicity increase, ground tilts, etc. | Pohai (NE China) | 1966–75 | 110 km per year | Scholz (1977) |
| Ground tilts | South Kanto (Japan) | 1950–70 | 20 km per year (W)* | Yamada (1973); Kasahara (1973a,b) |
| | Peru | 1966–70 | 60–70 km per year (NNE)* | Tanaka, Otsuka & Lazo (1977) |
| Ground strains | Tohoku (Japan) | 1968–75 | 40 km per year (N50 W)* | RGCM** (1977) |
| | Tohoku (Japan) | 1968–75 | 19 km per year (W)* | RGCM** (1977) |
| Postseismic fault slip | White Wolf (Calif.) | 1957–74 | 7–16 km per year | Stein, Thatcher & Castle (1977) |

*Direction of migration.
**Research Group for Crustal Movement.
***Order of magnitude estimate.

enon. Following the first discovery of migrating tilt anomalies in the South Kanto district, Japan (Yamada, 1973; Kasahara, 1973*a,b*), examples of a similar kind have been recognized in several other areas, such as the Tohoku district, Japan (Ishii, 1977), the West Cordillera Mountains, Peru (Tanaka, Otsuka & Lazo, 1977) and the White Wolf fault, California (Stein, Thatcher & Castle, 1977).

Fig. 7.22 shows the observed maximum shear strain accumulation at five stations in the Tohoku district, Japan. Excellent correlation is noticed among the first three stations on the Pacific side, where the events arrive progressively at SNR, MYK and HMK, in that order. This suggests northwestward migration. More precisely, a migration velocity

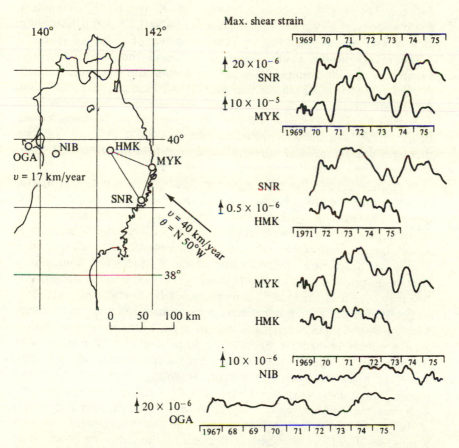

Fig. 7.22. Changes in maximum shear strain at five stations in the Tohoku district, Japan. Time axes are adjusted between the pairs of stations in order to plainly illustrate their correlation. Time scales for the repeated traces (SNR, MYK, HMK) have been eliminated. The index map to the left shows the location of the stations. (From Ishii, 1977.)

of 40 km per year in the direction N50°W was found from the phase differences at these stations. The other two stations on the Japan Sea side gave a velocity of 19 km per year (westward), although the direction could not be determined with certainty since a third station was unavailable in this group. The phase relation between the Pacific and the Japan Sea groups is also uncertain, but we may provisionally accept a general trend of westward migration across the Tohoku district (see the phase relations of the records at the five stations compared at the right of fig. 7.22).

The wave-form is strikingly similar between adjacent stations (cf. SNR and MYK), but it becomes less similar for remote stations (cf. SNR and OGA). In other words, migration seems dispersive and dissipative with distance. The wave-form change between MYK and HMK indicates a phase velocity of 40 km per year at periods of about 5.3 years, decreasing quickly with period to 20 km per year at periods of about 5.8 years.

The events in the South Kanto district and the West Cordillera Mountains provide examples of migrating ground tilts (see table 7.6) and showed velocities of 20 km per year and 60 to 70 km per year. Note that the events in the Tohoku and Kanto districts are characterized by migration from east to west, whereas the event in Peru is characterized by migration from south to north (see table 7.6). These examples seem to suggest a general trend of migration landward from the ocean (cf. §7.3).

*Models.* The observed migration velocities, as mentioned above, are lower than the ordinary elastic sound velocities in the crust by several orders of magnitude. Without doubt, some non-elastic properties of the medium play an essential role in these events. To construct a theoretical model, one may introduce a thin layer of non-elastic material, such as a fault gouge or fracture zone, between two elastic blocks which undergo a relative movement, i.e. fault slip (Savage, 1971; Ida, 1974). Alternatively, one may take a layered medium as proposed by Bott & Dean (1973).

The Bott–Dean model is basically a two-dimensional layered system, with an elastic plate overlying a viscous layer which in turn overlies a semi-infinite rigid medium representing, respectively, the lithosphere and asthenosphere between a trench and a fixed continental block (fig. 7.23). Let $E$, $h_1$, $\eta$, $h_2$ and $k$ denote the Young's modulus and thickness of the elastic plate, the Newtonian viscosity and thickness of the viscous layer, and the wave number, respectively. Then horizontal displacement $u(x, t)$ of the plate in the direction of the $x$-axis, due to the horizontal forces on a small length of the elastic plate, is given by the following equation of motion

$$\partial^2 u/\partial x^2 = \sigma(\partial u/\partial t), \tag{7.6}$$

where inertial forces are neglected and $\sigma = \eta/h_1 h_2 E$ for an assumed linear

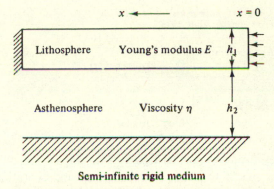

Fig. 7.23. Two-dimensional model of the lithosphere overlying the astheno-sphere. (After Bott & Dean, 1973.)

velocity–depth distribution across the viscous layer (for an alternative type of distribution, see Bott & Dean, 1973). We may apply (7.6) to the propagation of a horizontal shear stress by replacing $E$ by the rigidity, $\mu$. Bott & Dean compared (7.6) to the equations of one-dimensional heat conduction and obtained the following expressions for the response of an elastic plate of semi-infinite length to a sinusoidal boundary pressure $P = P_0 \sin \omega t$ of period $T = 2\pi/\omega$ applied at the end $x = 0$. The displacement is given by

$$u(x, t) = -(P_0/\sqrt{2}kE)e^{-kx} \cos (\omega t - kx + \pi/4), \qquad (7.7)$$

and the pressure is given by

$$P(x, t) = P_0 \, e^{-kx} \sin (\omega t - kx), \qquad (7.8)$$

where $k = \sqrt{(\frac{1}{2}\omega\sigma)}$. Let $v = \omega/k$, representing the velocity of a train of stress (or strain) in the $x$-direction, and defining the penetration $X$ as the distance to the point where the amplitude falls off to $e^{-1}$ of its value at $x = 0$, we have,

$$X = (ETh_1h_2/\pi\eta)^{1/2}, \quad \text{and} \quad v = 2(\pi Eh_1h_2/\eta T)^{1/2}. \qquad (7.9)$$

Bott & Dean assumed $h_1 = 80$ km, $h_2 = 250$ km, $\eta = 2 \times 10^{21}$ P and $E = 10^{12}$ dyn cm$^{-2}$, and calculated representative values of $X$ and $v$, to indicate the order of magnitude of penetration and velocity, respectively (the calculated values are given in table 7.7).

*Further implications.* A landward trend of migration of crustal deformations may lead us to locate the origins of the deformations about trench areas. Fig. 7.24 is a hypothetical view of a deformation front migrating from the ocean, where uniform migration velocities, as observed inland, are assumed (Kasahara, 1979). The event in the South Kanto district seems to have started propagating away from the junction of the Japan and Izu-Bonin Trenches in the early 1950s, and the event in the

Table 7.7. *Representative values of model para-*
*meters. (From Bott & Dean, 1973.)*

| Period, $T$ (years) | Penetration, $X$ (km) | Velocity, $v$ (km per year) |
|---|---|---|
| 1 | 10 | 62.8 |
| $10^2$ | $10^2$ | 6.28 |
| $10^4$ | $10^3$ | 0.628 |
| $10^6$ | $10^4$ | 0.0628 |

Tohoku district, from the northern part of the Japan Trench in the late 1960s. We may postulate either seismic or aseismic excitation, in these cases. If the former is the case, the 1953 and 1968 earthquakes off the

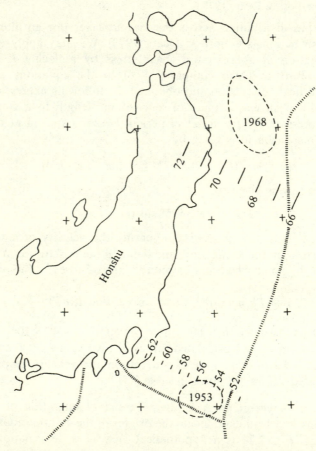

Fig. 7.24. Hypothetical picture of deformation fronts migrating landward from trench areas. Broken lines illustrate the Boso-oki (1953) and the Tokachi-oki (1968) earthquakes which might have caused the observed deformations. (From Kasahara, 1979.)

coast of Honshu are the most likely origins of the events in the South Kanto and Tohoku districts, respectively, and the stations will thereafter record only insignificant disturbances until the next major earthquakes occur. If, on the other hand, the stations continue to record significant disturbances in the future, we must reject the first possibility and accept the concept of aseismic excitation, which indicates that plate motion is irregular and intermittent even in an interseismic period.

Anderson (1975) modified the Bott–Dean model to give the more realistic model shown in fig. 7.25, which he used to explain the migration pattern of large earthquakes along an arc. He proposed that plate motions accelerate after great decoupling earthquakes and that most of the observed plate motions occur during short periods of time, separated by periods of relative quiescence. These conclusions are extremely interesting in view of the observational data reported by Mogi (1968c) on the acceleration trends of earthquake sequences (fig. 7.18), and by Utsu (1976b) on relatively long quiescence periods in seismic cycles (see §7.2.2).

Migration of crustal deformation and seismic activity, as observed in Northeast China preceding the Haicheng earthquakes (1975, $M = 7.3$), has proved its usefulness for earthquake prediction (§8.3.1). Scholz (1977), in a study of the Chinese data, observed a northwestward migration with a velocity of about 110 km per year.

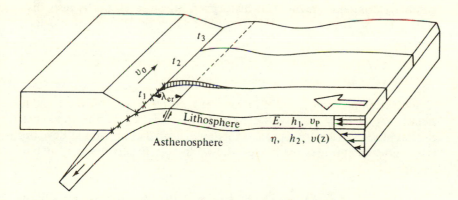

Fig. 7.25. The Anderson model. The oceanic lithosphere, of thickness $h_1$ and modulus $E$ under stress $\sigma_x$ is assumed to be sliding on an asthenosphere of viscocity $\eta$ and thickness $h_2$. The flow patterns are indicated schematically by the arrows to the right. A decoupling earthquake is indicated by the crosses, and $\lambda_{er}$ is the distance between the decoupling earthquake and the point of maximum tension. The decoupling earthquake occurs at time $t_1$ decreasing stresses in the lithosphere closest to the viewer and increasing stresses at the adjacent segments of the arc boundary, leading to subsequent earthquakes at times $t_2$ and $t_3$. The velocity of propagation of the stress wave is denoted $v_0$. (From Anderson, 1975. Copyright 1975 by the American Association for the Advancement of Science.)

# 8 Earthquake prediction

### 8.1.1 *The way to earthquake prediction*

The extensive classical literature on earthquake prediction illustrates the effort invested over many years to mitigate the worst effects of catastrophic earthquakes. Progress towards this goal has been accelerated over the last ten years by the promotion of systematic research, without which our way to the future would be uneven as it has been in the past.

Research on prediction is still at a stage in which empiricism plays an important role. Therefore, documentation of past events is essential to our understanding. This chapter does not aim to sketch prediction work historically. Readers interested in this topic should refer to the monograph by Rikitake (1976). The history of Japanese events is also discussed in the review by the Earthquake Disaster Prevention Society (1977).

### 8.1.2 *Structures of earthquake prediction*

A successful prediction must include the correct assessment of three elementary factors, namely the time, place and size of the predicted earthquake. In order to approach this goal, a prediction should develop systematically through the stages shown in fig. 8.1 discussed in detail below.

*Statistical prediction.* The first stage, statistical prediction has been discussed since the early days of prediction work. This type of prediction is based on the assumption that the earthquakes in a sequence in a particular region occur with a statistical character which does not change with time. If the sequence has either a predominant period or a probable correlation to some known external factor, the future activity may be predicted statistically (§8.2.2). This may be done relatively simply, but, even in the most straightforward cases, the prediction involves considerable uncertainty. Accordingly, statistical prediction is adequate for pre-

194

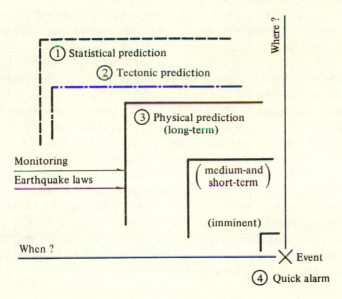

Fig. 8.1. Structure of an earthquake prediction. The size of the rectangles in the diagram represents uncertainties in the different types of prediction.

liminary work, such as the long-term planning of optimum monitoring systems and for disaster prevention work, in which the associated uncertainty is permissible.

*Tectonic prediction.* The second stage, tectonic prediction, is primarily concerned with the magnitude, type, and other tectonic parameters of an earthquake that is likely to occur in a given locality. The accumulation law of seismotectonic movement (§7.2.1) is the principal working hypothesis in this prediction technique. Precise time information is not the primary concern, but the lapse time in the seismic cycle, as discussed in §7.2.2 and §8.2.1, may be estimated.

*Physical (or precise) prediction.* The third stage aims at the precise determination of all three factors (time, place, size), by the recognition of meaningful seismic precursors to major earthquakes. Its ultimate task, therefore, is *deterministic prediction* (see the discussion of dilatancy models in §8.3.2). The essential requirements for this kind of work are a sufficient knowledge of the physical laws of earthquake processes and precise monitoring of physical states in and around the seismic region at the appropriate time.

*Quick alarm.* The last stage is not strictly prediction at all. However, it should be mentioned as a promising method of practical disaster reduction. This stage relies upon a fully-automated monitoring system, which is linked to critical public and/or industrial facilities so that they may be taken over promptly in an emergency. If our system detects and recognizes an imminent seismic event, the critical facilities could be controlled for a few seconds, a short but useful time, before the main seismic disturbance strikes.

Practical earthquake prediction will be carried out by working through three or four successive stages. A recent example was the successful prediction of the Haicheng earthquake, northeast China (1975, $M = 7.3$). Rikitake (1976) proposed an alternative classification of the stages of prediction by their respective time ranges, namely, predictions of the *long-*, *medium-* and *short-term* in addition to the *preliminary* one based on statistical prediction. *Extremely short range prediction* is also added to deal with imminent precursors.

Earthquake prediction is related to further problems, such as the reduction of earthquake hazards, public policy and earthquake control, which are discussed briefly in §8.5.

## 8.2    STATISTICAL AND TECTONIC PREDICTIONS

### 8.2.1    *Regionalization*

*Seismicity.* The purpose of regionalization is to study the geographical distribution of *earthquake risk* in a district or a country. We require sufficient information on the expected size and frequency of earthquakes for estimation of earthquake risk, but precise time information is not a matter of primary concern. In practice, we specify a certain time window for the future and estimate how much seismic energy (or, acceleration, intensity, etc.) per unit area will be registered during the specified future time period.

This may be done, to some extent, on the basis of seismic observations. Historical records, which allow past seismic activity to be inferred, are particularly useful for the determination of mean seismicity. Kawasumi (1951) examined historical earthquakes in Japan and compiled a seismic risk map of Japan (fig. 8.2). He first assigned each point on the map a frequency $n(I)$ of seismic intensity $I$, which had been registered at that point over the past $T$ years. Then, the maximum intensity $I_0$, expected over the next $t$ years, is approximately obtained in such a way as to satisfy the following relation

$$(t/T) \int_{I_0}^{\infty} n(I)\mathrm{d}I = 1 \tag{8.1}$$

where $I$ is conventionally assumed to be continuous. Kawasumi then converted the calculated $I_0$ into its equivalent acceleration to draw the map of the maximum seismic acceleration expected over the next one hundred years.

*Active faults*. Recurrence intervals appear to be especially long for inland faults, sometimes extending to thousands of years (§7.2.2; §8.2.3). The time base of available seismicity data, even that from historical records, is too short to represent the overall picture accurately. In fact, the Niigata earthquake, Honshu (1964, $M = 7.5$) occurred in an area that looked almost aseismic on the seismic risk map compiled in 1951 (fig. 8.2). Information from active fault studies is an indispensable complement to the available seismicity data.

An example of an active fault map may be seen in fig. 7.17. A serious problem in this work is the difficulty of fault recognition, particularly of submarine faults and those buried under thick alluvium. The evaluation of earthquake prediction, discussed in §8.42, illustrates this difficulty.

Wallace (1970) developed a classification of faults based on tectonic factors and assigned each segment of the San Andreas fault to a specific category of tectonic behaviour (fig. 8.3). The compound illustration of such information is useful for earthquake prediction and hazzards

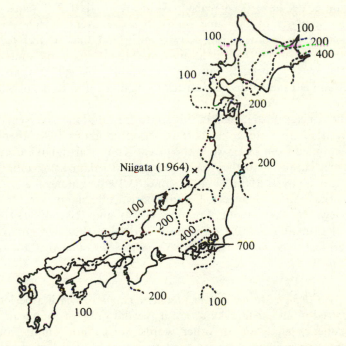

Fig. 8.2. Seismic risk map of Japan. The contour lines (dashed) show maximum ground acceleration in cm s$^{-2}$ expected in the next hundred years. (After Kawasumi, 1951.)

reduction. Consequently, various types of maps, such as seismotectonic maps (compilation of seismic, geodetic and geological information) and seismic zoning maps (compilation of seismic and earthquake engineering information) have been published in several countries (see Allen *et al.*, 1965; for a seismicity map of southern California).

### 8.2.2    *Prediction of earthquake sequences*

*Simple statistics.* The statistical approach to earthquake prediction has been studied for many years with special emphasis on the periodicity of earthquake sequences and the correlation between earthquakes and other phenomena.

Kawasumi (1970), for example, analysed historical documents on strong earthquakes felt at Kamakura, a city about 50 km south of Tokyo, and noticed a 69-year periodicity for strong earthquakes in the South Kanto district, which includes the Tokyo Metropolitan area. The city of Kamakura was the seat of the Japanese government around A.D. 1200, so that the seismic sequence there was fairly well documented. Thirty-two events of intensity V or higher on the Japan Meteorological Agency (JMA) intensity scale (about intensity VIII in the modified Mercalli scale) have been registered during the last 750 years. After a Fourier analysis of the time series of these events (each event was represented by a unit pulse), Kawasumi found that several periods in the power spectrum were predominant. Most remarkable among them was the 69-year period, with an uncertainty of $\pm 13.2$ years, which Kawasumi considered to be statistically significant in terms of Schuster's criterion.

If the conclusions reached in this study are valid, the Tokyo and South Kanto areas will enter a 'dangerous period' in about 1980, after $69 - 13$ years counted from 1923, when the Great Kanto earthquake occurred in this area. However, further examination of the data using more careful statistical analysis (Usami & Hisamoto, 1970; Shimazaki, 1972) has shown that the 69-year period is not statistically significant.

Reviews by Aki (1956), Lomnitz (1974) and Rikitake (1976) have show that many statistical predictions of earthquakes were unacceptable after critical tests. These articles also give reviews of correlation analyses.

*Earthquake recurrence in subduction zones.* Even if a periodicity is recognized in an earthquake series, it is valid only for the mean state of the sequence analysed. In other words, we cannot simply count the particular interval to the next earthquake taking the last event as the time origin. When prediction by simple statistics is used, special care

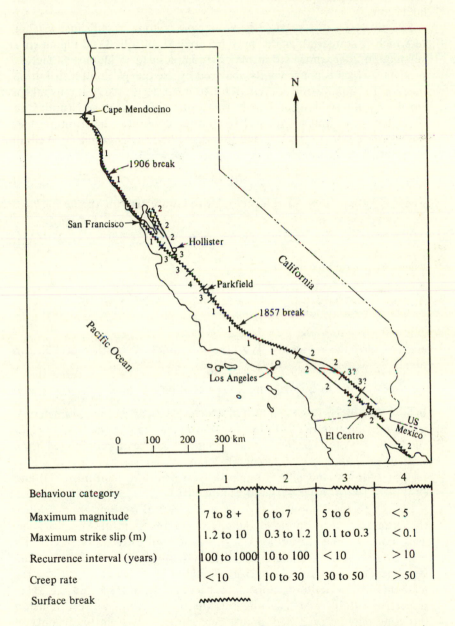

| Behaviour category | 1 | 2 | 3 | 4 |
|---|---|---|---|---|
| Maximum magnitude | 7 to 8 + | 6 to 7 | 5 to 6 | < 5 |
| Maximum strike slip (m) | 1.2 to 10 | 0.3 to 1.2 | 0.1 to 0.3 | < 0.1 |
| Recurrence interval (years) | 100 to 1000 | 10 to 100 | < 10 | > 10 |
| Creep rate | < 10 | 10 to 30 | 30 to 50 | > 50 |

Surface break

Fig. 8.3. Classification of the San Andreas fault into four behaviour categories. Creep rate in the category list represents the percentage of secular offset attributable to creeping in the respective segment. (From Wallace, 1970.)

must be taken not to extend the application of the statistics beyond its limitations.

Seismic cycles in a subduction zone have tectonic significance (§7.2.2), to which a statistical model may be applied as follows. Slipping in a subduction zone under strain may be compared to a life test of factory products, which is conveniently analysed by use of a Weibull distribution function for quality control research (Rikitake, 1976, 1977). Let us take a small time interval $\Delta t$, and write the probability for crustal rupture to occur between $t$ and $t + \Delta t$ as $\lambda(t)\,\Delta t$, on the condition that rupture does not occur prior to $t$. The hazard rate, $\lambda(t)$, is given in a Weibull distribution by

$$\lambda(t) = Kt^{m}, \tag{8.2}$$

where $K > 0$ and $m > -1$. The cumulative failure rate for the period $t = 0$ to $t$ is expressed as

$$F(t) = 1 - R(t), \tag{8.3}$$

where $R(t)$ is defined by

$$R(t) = \exp\left[ -\int_{0}^{t} \lambda(t)\mathrm{d}t \right] = \exp\left[ -Kt^{m+1}/(m+1) \right],$$

$$\tag{8.4}$$

and is called the reliability. We take the double logarithm of $1/R$ and obtain:

$$\ln\,\ln(1/R) = \ln[K/(m+1)] + (m+1)\,\ln\,t; \tag{8.5}$$

$\ln \ln (1/R)$ is linearly related to $\ln t$, which suggests that the parameters $K$ and $m$ may be empirically obtained by fitting a straight line to $\ln \ln (1/R)$ versus $\ln t$ plots. The $R$ versus $t$ relation can be approximated from experimental data (i.e. a histogram of observed seismic intervals in a region). Once $K$ and $m$ are obtained, we substitute them into (8.3) and (8.4) in order to revise $R(t)$ and obtain $F(t)$.

Fig. 8.4 shows theoretical $F(t)$ curves for several tectonic zones. Notice the extremely steep onsets of the curves for the Central America zone. This indicates that earthquakes in this zone tend to recur very regularly with a fairly short seismic interval. In fact, its mean return period and standard deviation, calculated from $K$ and $m$ (see Rikitake, 1976, for the mathematical procedure), appear to be only about 34.5 and 3.6 years respectively (table 8.1). This suggests that the earthquake recurrence in this zone may be predicted almost deterministically if an uncertainty of several years is tolerable. The Nankai–Tokai zone, off the Pacific coast of Honshu, on the other hand, is characterized by a much longer period and larger standard deviation so that statistical prediction is more difficult.

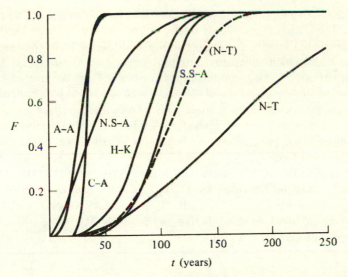

Fig. 8.4. Cumulative probabilities of great earthquakes recurring in the following zones: N–T, Nankai–Tokai zone (all data); (N–T), Nankai–Tokai zone for the data set after 1400; H–K, Hokkaido–Kurile zone; A–A, Aleutian–Alaska zone; C–A, Central America zone; N.S–A, northern South America zone; S.S–A, southern South America zone. (From Rikitake, 1977.)

Table 8.1. *Mean return period and its standard deviation for various tectonic zones. The values in parentheses for the Nankai–Tokai zone are obtained from the analysis of the data for the last 600 years. (From Rikitake, 1977.)*

| Zone | Mean return period (year) | Standard deviation (year) |
|---|---|---|
| Nankai–Tokai (Japan) | 170 | 68.9 |
| (Nankai–Tokai) | (117) | (35.0) |
| Hokkaido–Kurile | 85.3 | 24.6 |
| Aleutian–Alaska | 27.2 | 8.9 |
| Central America | 34.5 | 3.6 |
| Northern South America | 46.3 | 30.0 |
| Southern South America | 100 | 22.5 |

### 8.2.3 Prediction of seismotectonic parameters

Studies of active faults with special reference to their coseismic behaviour has led us to introduce the accumulation law of seismotectonic movements given in §7.2.1. Therefore, we can predict the possible size (magnitude) and mode of a future seismic event in an area provided its tectonic environment is sufficiently well known.

The most fundamental information, such as the location, orientation, length, slip type and slip direction of a potential seismic source (i.e. the active fault itself) may be obtained from a map such as fig. 7.17*a* or *b*. Then, the magnitude *M* of this event may be estimated either from the predicted length, *L*, of the coming faulting, or from the interseismic lapse time which measures potential energy accumulated to that moment. *M* and *U* (slip amplitude) are obtained by the use of empirical formulae such as log $L = aM - b$ and log $U = cM - d$ (see appendix 2 for details). We usually approximate *L* by a traceable length of the geological fault under consideration. If the fault undergoes only fractional breaking, therefore, these parameters will be observed to be smaller than predicted (Matsuda, 1976).

If we also know the accumulation rate, *S* (see fig. 7.14), the recurrence time, *T*, may be obtained by (7.4). Table 8.2 compares these tectonic parameters for several active faults. It is necessary to modify (7.4) slightly in order to deal with an active fault which has an associated creep component (Wallace, 1970).

### 8.3 PHYSICAL PREDICTION

#### 8.3.1    *Basic considerations*

Successful physical prediction of earthquakes must be founded, directly or indirectly, on an improved understanding of the nature of earthquakes. The following topics may be especially important and are discussed briefly for future reference.

*Accumulation law of seismotectonic movements.* Most fundamental to the study of earthquake prediction is a realization that the present seismotectonic activity is primarily a recent development in the crustal dynamics although the origin time of activity extends back hundreds of thousands of years. From this we assume that a future slip on a fault will be similar to slips in past events. The prediction of seismotectonic parameters in a future event is discussed in §8.2.1.

*Ultimate strain.* Data on the ultimate strain of the Earth's crust are also essential for prediction. The strain level of about $10^{-4}$ has been widely accepted as ultimate from various sorts of evidence (§2.3.4). Recently, Rikitake (1975*a*) statistically analysed 26 levelling and triangulation data over earthquake areas and deduced that the value of $4.7 \times 10^{-5}$ is a more likely mean strain at rupture (fig. 8.5). In both cases, however, the suggested value should be taken as an order-of-magnitude estimate only.

*Seismic gap.* The concept of a seismic gap is a useful working hypothesis for prediction purposes, as shown by the example of the 1973

Table 8.2. *Tectonic parameters of active faults. The magnitudes of earthquakes given in parentheses are calculated values. (From Matsuda, 1976.)*

| Fault | Displacement rate (mm per year) | Related earthquake | Magnitude, M | Recurrence time, T (years) | Reference |
|---|---|---|---|---|---|
| Median Tectonic Line (Shikoku, Japan) | 5–10 | | (8) | 1000–3300 | Okada (1973), Matsuda (1975a) |
| Atera (Honshu, Japan) | 5.3 | | (8) | 300 | Matsuda (1975a) |
| Tanna (Honshu, Japan) | 2 | Kita-Izu (1930) | 7.0 | 1000 | Kuno (1962) |
| Irozaki (Honshu, Japan) | 0.1–1 | Izu-Hanto-Oki (1974) | 6.9 | 1000 | Matsuda (1975b) |
| San Andreas (Calif., USA) | 13 | San Francisco (1906) | 8.25 | 700 | Wallace (1968) |
| Wairau (North Is., NZ) | 3.5 | | (8) | 100 / 500–900 | Wallace (1970), Lensen (1968) |
| Coyote Creek (Calif., USA) | 3 | Borrego Mt (1968) | 6.4 | 200 | Clark, Grantz & Rubin (1972) |

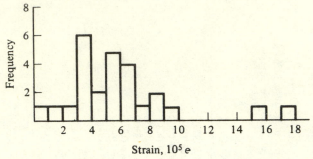

Fig. 8.5. Histogram of the ultimate strain, $\varepsilon$, of the Earth's crust deduced from levellings and triangulations over earthquake areas. (From Rikitake, 1975a.)

Nemuro-oki earthquake (§7.2.3). This idea originally applied to a considerable area of seismic quiescence in an inter-plate seismic plane. However, this term is sometimes used in a broader sense, to represent a local intra-plate quiescence, of a temporary ceasing of microseismicity in a seismic region prior to a principal earthquake. Therefore, the term must always be clearly defined when it is used to avoid unnecessary confusion in evaluation of data.

*Migration of crustal activity.* Migration of crustal activity (§7.3.2) may also be a promising working hypothesis for prediction. The successful prediction of the 1975 Haicheng earthquake in the People's Republic of China used this hypothesis. A sequence of major earthquakes in northeast China since the late 1960s led Chinese seismologists to suspect a trend of northeastward migration of seismic activity across the Pohai Bay. Taking this apparent trend into consideration, they intensified their fieldwork in the suspicious district so that they may give an accurate warning of the earthquake (Chu, 1976; Raleigh *et al.*, 1977). Later, Scholz (1977) reviewed their successful prediction and explained its background on the basis of a hypothetical 'deformation front' that propagated for 1000 km through northeast China at a velocity of about 110 km per year (fig. 8.6). The Anatolian earthquakes (fig. 7.21) and several other examples of earthquake migration suggest this technique to be a useful tool for long-term prediction.

### 8.3.2 *Clues to precise prediction*

*Preseismic land movements.* Evidence for preseismic land movements, for example, anomalous sea-level changes prior to seismic events, has been documented for many centuries (Imamura, 1937; Earthquake Disaster Prevention Society, 1977). Recently, further examples of preseismic land movement have been collected by instrumental methods

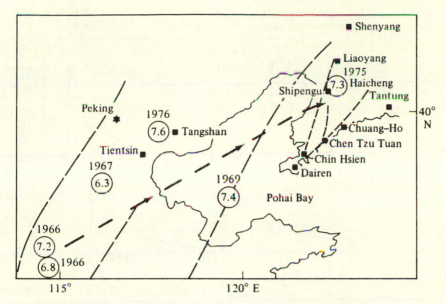

Fig. 8.6. Map of northeast China, showing all earthquakes larger than $M = 6$ since 1966 and the path of a hypothetical deformation front. (From Scholz, 1977.)

using levelling techniques, tide-gauges, tiltmeters, strainmeters, etc. (see tables 15-XIII and 15-XIV in Rikitake, 1976).

Fig. 8.7 shows the preseismic land movements in the 1964 Niigata earthquake ($M = 7.5$), revealed by: tide-gauge observation at Nezugaseki (Tsubokawa, Ogawa & Hayashi, 1964); tiltmeter observation at Maze (Kasahara, 1973); and levelling surveys, repeated along the eastern side of the epicentral area (Tsubokawa, Ogawa & Hayashi, 1964; Dambara, 1973). The accumulation of anomalous land movement since about 1955 is equally well shown by the three methods. The results correlate well, in spite of the different methods used and the fact that the stations are spread over a distance of 100 km.

Note the direction of movement, which reversed from uplift to subsidence shortly before the earthquake occurrence. This reversal of polarity (or remarkable acceleration or deceleration in a broader sense) is retrospectively considered to be an indication of an imminent catastrophe.

Fig. 8.7 shows that the preseismic uplift started over the entire length of the future epicentral region (about 100 km) about 10 years prior to the earthquake. It has been determined empirically that the larger the earthquake magnitude, the longer the *precursor time*, i.e. duration time of a preseismic event prior to earthquake occurrence. A general rule is not yet known for the size of a precursor area. However, we may provisionally assume that it is comparable with the area of future coseismic land deformation (§2.3.2), where the major strain processes take place.

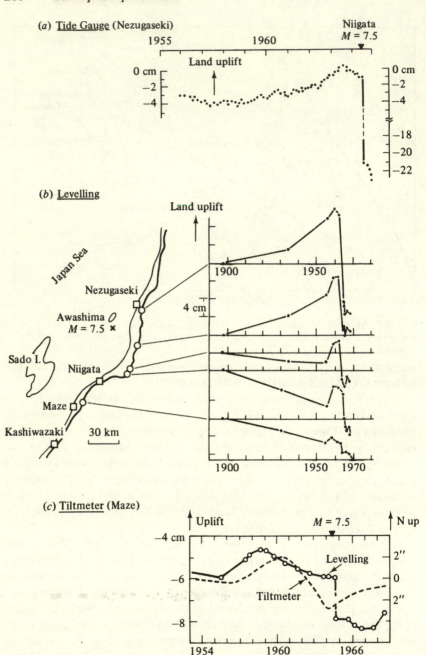

Fig. 8.7. Preseismic land movements in the Niigata earthquake, Honshu (1964, $M = 7.5$), as observed by: (*a*) tide-gauge, (*b*) levelling; and (*c*) tiltmeter. (Compiled from Tsubokawa, Ogawa & Hayashi, 1964; Dambara, 1973; and Kasahara, 1973.)

*Foreshocks.* Many documents, historical and modern, describe precursory activity of small earthquakes in or around a future source region. Laboratory experiments support the possibility that numerous microcracks in a rock test piece may lead to a main fracture. Fig. 8.8 shows a foreshock sequence in the Haicheng area. Notice the marked increase in local seismicity about one day before the onset of the main earthquake, which was followed by a short-term inactivity of several hours prior to the main event. This trend of short-term inactivity may be a promising clue for deterministic prediction, provided that its behaviour is fully understood.

Occurrence of foreshocks are fairly easily recognized if the area of concern has long been inactive, as was the case for the Haicheng earthquake. In a seismically 'noisy' area, on the other hand, discrimination of foreshocks from other false events (e.g. swarm activity, followed by no principal shock) generally proves troublesome.

In order to discriminate foreshocks from the flase events, we sometimes test the magnitude–frequency relation of a sequence (§2.4.1), using the hypothesis that the $b$-value (gradient constant in (2.30)) for foreshocks tends to appear lower than the normal value (0.8–1.0) (see also Hsu, 1976 for further discrimination methods).

Successful discrimination of foreshocks is urgently needed, since a considerable number of disastrous earthquakes, though not all (see, for example, table 8.5 for statistics), are likely to be preceded by foreshocks.

*Geophysical precursors.* Preseismic anomalies have been noticed in the study of many other geophysical phenomena, in addition to the

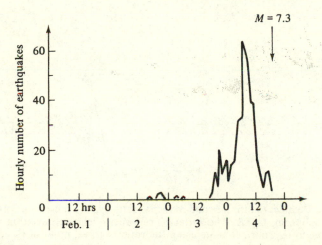

Fig. 8.8. Sequence of foreshocks before the Haicheng earthquake, northeast China (1975, $M = 7.3$; Shihpengyu Seismic Observatory). (After Hsu, 1976.)

extraordinary behaviour of fish and animals described in many legends. Tables 15-XIII and 15-XIV in Rikitake (1976), for example, compile an extensive set of precursors experienced at various times and in various regions of the world of the following types:

land deformation;
tilt and strain;
foreshock;
*b*-value;
microseismicity;
source mechanism;
fault creep anomaly;
$v_P/v_S$;
$v_P$ and $v_S$;
geomagnetism;
Earth currents;
resistivity;
radon;
underground water;
oil flow.

Precursory changes in seismic wave velocities ($v_P$ or $v_S$) or velocity ratio ($v_P/v_S$) in a seismic region have been noticed in the last few years, especially since the detailed work of Russian scientists in Garm, USSR. Fig. 8.9 demonstrates the temporal changes in seismic wave propagation which preceded strong earthquakes in Garm. The ordinate represents

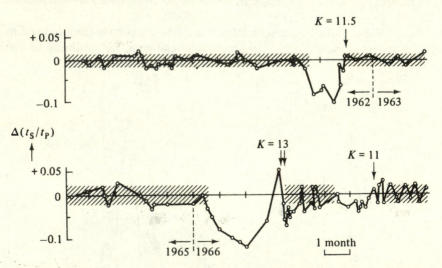

Fig. 8.9. Changes in the seismic velocity ratio, $v_P/v_S$, preceding strong earthquakes in Garm, USSR. The ordinate represents relative deviation of velocity ratio $v_p/v_s$ from a normal value using the ratio of travel times of P- and S-waves ($t_P$ and $t_S$ respectively) from a nearby focus across the region tested. $K$ is the Russian energy index of earthquakes. (After Semenov, 1969.)

relative deviation of velocity ratio ($v_P/v_S$) from a normal value, and $K$ denotes the Russian energy index of earthquakes (see §2.1), which is given by the common logarithm of seismic energy in joules (e.g. $K = 13$ corresponds to seismic energy of $10^{13}$ J, i.e. $10^{20}$ erg).

Preseismic changes in velocity ratios are remarkable in this example, first decreasing, then returning to the normal level when the main shock occurs. If this behaviour is universal, it will provide a useful signal for deterministic prediction. Extensive work has been conducted in several countries to confirm this possibility. The results are rather confusing, however, as some studies confirm the possibility while others do not (see for example Whitcomb, Garmany & Anderson, 1973; Kanamori & Chung, 1974; Aggarwal *et al.*, 1975). The confusion may be accounted for partly by the insufficient precision of conventional seismometrical data, and partly by rockmechanical anisotropy. That is, the magnitude of this effect may appear different, depending on the direction of wave path relative to the predominant orientation of microcracks.

As laboratory experiments have shown, the phenomenon of velocity decrease is expected to precede a main fracture. It would be a great advantage if the velocity measurement techniques could be improved to enable precursory cracking at a particular depth to be accurately monitored.

The possible use of many other geophysical precursors for prediction has been intensively discussed (see for example Wyss, 1975; Rikitake, 1976; Raleigh *et al.*, 1977; Suzuki & Kisslinger, 1977).

*Precursor time.* Tsubokawa (1969, 1973) compiled the Japanese data on preseismic land movements and discovered a linear relation between the logarithmic precursor time $T$ and the magnitude $M$ of a forthcoming earthquake, i.e.

$$\log T = 0.79\ M - 1.88, \tag{8.6}$$

where $T$ is measured in days. Similar empirical formulae have been proposed by several researchers and are compared in fig. 8.10. The formulae are:

$$\log T = 0.685\ M - 1.57 \text{ (Scholz, Sykes \& Aggarwal, 1973)}, \tag{8.7}$$

$$\log T = 0.80\ M - 1.92 \text{ (Whitcomb, Garmany \& Anderson, 1973)}, \tag{8.8}$$

$$\log T = 0.76\ M - 1.83 \text{ (Rikitake, 1975}b), \tag{8.9}$$

in which $T$ is measured in days, as in (8.6). Rikitake noticed that the precursors so far reported include a small number of special data which do not follow (8.9). This type of precursor tends to occur frequently before relatively large earthquakes, with almost uniform precursor time ($\log T \approx -1$, i.e. $T$ is several hours) for magnitudes larger than 5 (see

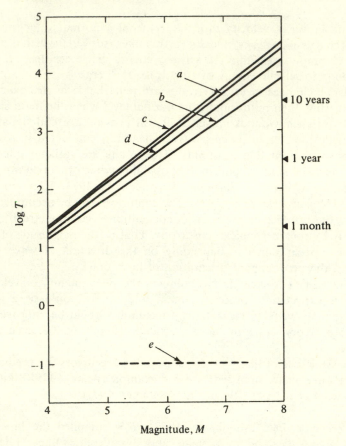

Fig. 8.10. Precursor time, *T* (measured in days) versus magnitude, *M*, relations by: (*a*) Tsubokawa (1969, 1973); (*b*) Scholz, Sykes & Aggawal (1973); (*c*) Whitcomb, Garmany & Anderson (1973); and (*d*) and (*e*) Rikitake (1975*b*).

broken line in fig. 8.10). Rikitake speculated that these special precursors may be closely correlated with the creep-like rupture that would occur at the focal region immediately before the main rupture. A physical explanation of (8.6) to (8.9) may be given by the dilatancy model discussed below.

*Dilatancy model*. Physical mechanisms of dilatancy in rocks have already been discussed in §6.3. As Scholz, Sykes & Aggarwal (1973) demonstrated (fig. 6.9), a dilatancy model (of wet type, in this case) may be used to explain many kinds of seismic precursors such as preseismic decrease and recovery of seismic velocities, decrease in electrical resistivity, increase of water flow rate (or of radon emission), land uplift, and activity and inactivity of foreshocks. All these geophysical anomalies have been observed to some degree in various seismic fields preceding earthquake occurrence. A merit of the dilatancy model is that a simple process of water diffusion into newly created pores in rocks explains a great variety of precursors in detail.

In support of their model, Scholz, Sykes & Aggarwal explained the precursor time–magnitude relation by fluid (or, water in this case) diffusion in a porous medium. They related the precursor time, $T$, to the longest dimension, $L$, of the aftershock zone of the respective earthquake by the formula,

$$T = L^2/c, \tag{8.10}$$

where $c$ is a constant which is related to parameters such as the permeability and the porosity of the rock, as well as the viscosity and compressibility of the pore fluid.

The dilatancy model in this form raises the basic question of whether water diffusion can occur, as Scholz, Sykes & Aggarwal hypothesized, at considerable depth (e.g. $\geqq 10$ km). A dry model (§6.3) seems better in this respect, although further tests are necessary on both models to recognize which is more acceptable.

## 8.4 THE PRACTICE OF EARTHQUAKE PREDICTION

### 8.4.1 *Monitoring*

The primary purpose of prediction observations is prompt recognition of anomalous crustal activities preceding an earthquake. To do this, we monitor the temporally changing (or unchanged) physical states in the seismic region of concern. Many review papers and monographs have discussed the instrumentational and technical problems of monitoring (see for example Tsuboi, Wadati & Hagiwara, 1962; Press *et al.*, 1965; Rikitake, 1976), which we will not repeat here.

Instruments used for prediction work do not differ much from those used for basic geophysical research. The main requirement of monitoring for prediction as opposed to other work is a large station network, equipped with a dense distribution of various instruments over a wide area. To facilitate efficient operation the large number of stations and instruments are usually integrated into a comprehensive system and computers used for quick data processing. Mobile observation parties are usually used to complement the fixed station network.

The monitoring systems now operated as national projects in several countries (e.g. Japan, USA, USSR, etc.) are similar although their detailed structure may differ from country to country. The system used in the People's Republic of China is unique in this respect. Their station network, which is similar to those in other countries, is systematically linked and complemented by a large number of teams of volunteers, who work on the various kinds of observations that professional scientists could not complete by themselves.


### 8.4.2    *Evaluation of earthquake prediction*

*Approaches to physical prediction of earthquakes.* Earthquake occurrence is basically understood as a process of strain accumulation in rocks, leading to a catastrophic fracture which radiates the accumulated strain energy as waves (§6.1). Therefore, the following two categories hold the key to successful prediction:

(*a*) to recognize that part of the crust has reached a critical state, i.e. that strain has reached its ultimate level;

(*b*) to recognize, by some decisive precursor(s), that preliminary cracking has started, prior to a main fracture.

The first category is concerned with a crustal condition which predisposes a particular region in the crust to earthquake occurrence. Repetition of earthquakes at a fault is subject to several seismotectonic laws as seen in §8.3.1, which enable us to tell whether a region is mechanically critical or not. This is, of course, only an approximate prediction, yet it can be useful for a long-term prediction if sufficient seismotectonic data are available.

The second category must be considered for more precise prediction, as seen in §8.3.2. As has already been pointed out on several occasions, fracturing is essentially a stochastic process. This implies that we can predict occurrence of an earthquake only in terms of a probability, but not deterministically. Nevertheless, we aim at deterministic prediction as the goal of our work. An excuse for this optimism is that we can transform the very difficult problem of precise determination of time into a relative problem. Suppose a dilatancy model applies to a seismic process, then our task may be reduced to a prediction of how soon a preliminary fracture will lead to a main fracture. We may call it *relative prediction*, in this sense. It is certainly easier than the absolute prediction of the time of occurrence, though there still remain difficult problems such as recognition of precursors and accurate estimates of a precursor time (§8.3.2).

*Evaluating earthquake predictions by 'scores'.* Let us take the second category as an example and study how to score earthquake prediction. There are basically two kinds of scores. They are the *truth rate* $p_1$ and the *alarm rate* $p_2$, where we simply define them as the frequency ratios of successful prediction to issued prediction, and of successful prediction to frequency of events to be predicted, respectively. In practice, we introduce a certain criterion $X$ for testing anomalies, and issue a prediction every time our observation meets this criterion. This is conventionally done by visual inspection of records, but for more automatic and objective detection of anomalies, computer processing of data is preferable. Ishii (1976), for example, tested strainmeter records using a Chebychev filter and identified 'anomalies' when the deviation of observation from the future normal value by extrapolation exceeded a certain threshold. Similar filters, e.g. the Wiener filter (Båth, 1974), may also be useful.

Recognition of successful prediction is a complex problem. For the sake of simplicity, therefore, let us assume reasonable tolerances for the predicted location, magnitude and time of an earthquake, and recognize the prediction as successful only when the three parameters of the occurred event all fall in the respective tolerance (criterion $Y$). Dependence of the scores on the criteria is complex, but, qualitatively, we can understand their mutual relations as follows. If the criterion $X$ is less specific, $p_2$ is improved at the expense of $p_1$. If $Y$ is less specific, on the other hand, $p_1$ is improved. Deceptive improvement of either $p_1$ or $p_2$ in this way will make the prediction meaningless, however (imagine everyday prediction in an area). For optimum design of prediction work, therefore, we must define and evaluate a merit factor of prediction (see below).

The previous example has dealt with prediction by a single-channel observation. If two or more independent observations are correlated, the scores of prediction may be significantly improved. Suppose, for example, we have two independent observations, A and B. Then, the previous rates are rewritten as shown in the following table (Utsu, 1977).

| Rate | Single observation | Double observation (A, B) | | |
| --- | --- | --- | --- | --- |
| | | Individual component | Recognized by A or B | Recognized by A and B |
| Truth rate | $p_1$ | $p_1(A), p_1(B)$ | $p_1(A\cup B)$ | $p_1(A\cap B)$ |
| Alarm rate | $p_2$ | $p_2(A), p_2(B)$ | $p_2(A\cup B)$ | $p_2(A\cap B)$ |

These components are interrelated as follows:

$$p_2(A\cap B) = p_2(A)p_2(B),$$

$$p_2(A\cup B) = p_2(A) + p_2(B) - p_2(A)p_2(B),$$

$$p_1(A\cap B) = \left[1 + R\left(\frac{1}{p_1(A)} - 1\right)\left(\frac{1}{p_1(B)} - 1\right)\right]^{-1}, \qquad (8.11)$$

$$p_1(A\cup B) = \frac{p_2(A\cup B)}{p_2(A)/p_1(A) + p_2(B)/p_1(B) - p_2(A)p_2(B)/p_1(A\cap B)},$$

and,

$$R = 1/[(1/p_0) - 1],$$

where $p_0$ denotes the truth rate by 'random' prediction. If $R$ is small enough in the third equation, $p_1(A \cap B)$ can be very high in comparison with both $p_1(A)$ and $p_1(B)$.

*Merit factor of earthquake prediction.* Scoring of earthquake prediction may be compared to that of a baseball game, where a batter is expected to swing his bat to make contact with every strike ball, and not to swing at any false ball. Let us compare a strike ball to a unit time interval (e.g. a day) in which earthquakes occur. Then, for successful prediction, an alarm must be given to each seismic interval, and no alarm must be given to the aseismic intervals. In this analogy, the alarm and truth rates refer to the correct recognitions of strike balls and the hitting scores, respectively. Let us take a sufficiently long period of observation time, in which sufficient numbers of anomalies and earthquake events occurred for statistical predictions to be significant. Suppose we have issued, during this period, $m$ true (successful) predictions and $n$ false predictions and have not issued predictions at all for the other $\mu$ earthquakes. Then, the truth and alarm rates are given respectively by

$$p_1 = m/F, \quad p_2 = m/M, \tag{8.12}$$

where $F = m + n$ and $M = m + \mu$ denote the total numbers of alarms and earthquakes to be predicted, respectively.

Utsu (1977) started from this idea and formulated the *merit factor of earthquake prediction*, $E$. In order to define this quantity, let us first introduce the following parameters:

   $l$ damage in an earthquake, when properly forecast,
   $L$ damage in an earthquake, when not forecast,
   $f$ costs of hazard reduction after a prediction,
   $C$ costs of prediction work (total amount during the period),

where in each case the average figure is estimated for a 'standard' seismic event. Then, total damage during the above-mentioned period will amount to $ML$ if we have no prediction work, and to $Ff + ml + \mu L + C$, if we have issued predictions. We now define the merit factor $E$ as the difference of damage in the above two cases relative to the one without prediction, so that we may write $E$, using (8.12), as:

$$\begin{aligned}
E &= (mL - ml - Ff - C)/ML \\
&= p_2[1 - (l/L) - (1/p_1)(f/L)] - c, \tag{8.13}
\end{aligned}$$

where $c = C/ML$. For perfect prediction $(p_2 = p_1 = 1)$, the merit factor becomes

$$E_0 = 1 - (l+f)/L - c, \qquad (8.14)$$

so that we obtain

$$-\infty < E < E_0 < 1. \qquad (8.15)$$

In practice, $p_1$ and $p_2$ are related to one another. In a cautious (pessimistic) prediction, we can improve $p_1(\rightarrow 1)$ at the expense of $p_2(\rightarrow 0)$. In an optimistic prediction, on the other hand, $p_2$ can approach 1, but $p_1$ will go down to zero at the same time. Qualitatively, therefore, the merit factor $E$ will be poor if we take a too pessimistic, or too optimistic, attitude in the prediction (Utsu, 1977).

*Predictable earthquakes.* The Chinese work on the Haicheng earthquake has shown that we really can predict an earthquake in some cases, under favourable conditions. Consequently, the question about whether prediction is possible or not should hereafter be raised in a more specific manner; for example, what types of earthquakes are predictable, and what percentage of the earthquakes to be predicted look predictable within the present prediction capability. Several sets of Japanese statistics, which are useful in the consideration of these problems, are given below.

Matsuda (1977) compiled the seismotectonic aspects of past disastrous earthquakes which occurred on the Japanese mainland and evaluated how many of them could have been predicted by geomorphic means (§8.2.1). Table 8.3 classifies these events into several categories. From this table we can see that 12 events, i.e. 52% of the sample

Table 8.3. *Categorization of earthquakes for potential prediction using active fault surveys. (After Matsuda, 1977.)*

| Disastrous earthquake of magnitude $M \geq 6.5$ occurring inland, Japan, 1868–1975 | | |
|---|---|---|
| Category | Frequency | Percentage of total |
| Earthquake with surface evidence of faulting | 10 | 40 |
| Earthquakes to which an active fault could be attributed | 12 | 52 |
| 'Tectonically' predictable earthquakes, if sufficient geomorphic data were provided | 8 | 35 |
| Total | 23 | 100 |

earthquakes of $M \geq 6.5$, could be attributed to recognizable active faults. However, only 8 of these 12 events (35% of the total) look predictable using presently available geomorphic techniques. In other words, the remaining fifteen of the 23 sampled events could not have been located by this method. This figure is worse if smaller earthquakes and off-coast earthquakes are also taken into consideration, since the surface evidence in these events is relatively small and rarely accumulates amplitudes large enough to be distinguished from erosion and other geological disturbances.

Table 8.4 classifies anomalous land uplifts discovered by repeated levellings, into several supposed types. Note that 9 of the 56 events are recognized as inland preseismic movements (the coastal movements related to off-coast major earthquakes are excluded from the table). For prediction work, signal discrimination from the postseismic, folding and compactional movements is relatively easy. However, the discrimination of the seismic swarm and unclassified categories, which numbered 27 events in total, is of great importance. If no further discrimination technique is available, therefore, only one quarter of the suspicious uplifts can be considered as seismic precursors, i.e. from the present data, the truth rate can be estimated as $\frac{1}{4}$.

Table 8.4. *Various aspects of 'anomalous' land uplifts in Japan, 1880–1976. (After Sato & Inouchi, 1977.)*

| Criterion | Seismic precursor | Post-seismic effect | Active folding, etc. | Seismic swarms | Alluvium compaction | Unclassified | Total |
|---|---|---|---|---|---|---|---|
| A | 1 | 0 | 3 | 2 | 0 | 3 | 9 |
| B | 3 | 2 | 4 | 0 | 2 | 7 | 18 |
| C | 5 | 3 | 4 | 0 | 2 | 15 | 29 |
| Totals | 9 | 5 | 11 | 2 | 4 | 25 | 56 |

A: significant (rate at 0.3 cm per year or higher; benchmarks well preserved)
B: fair (rated at 0.3 cm per year or higher; benchmarks faint or missing)
C: faint (rated at 0.2 cm per year or less)

The Kanto district, Honshu, Japan, frequently generates strong earthquakes of various types, such as swarm and foreshock types in addition to single shock and aftershock types. Table 8.5 shows the statistics of such events for different magnitude ranges. The period of sampling is not sufficiently long for statistically significant conclusions, but it does contain useful information for the present discussion. Suppose we take a simple hypothesis that a large earthquake will occur following a group of microearthquakes and on the basis of this hypothesis, issue a test prediction every time we observe a group of microearthquakes in this

Table 8.5 *Frequency of various types of earthquake sequence in the Kanto district, Honshu, June 1971 – December 1975. (After Kakimi et al., 1977.)*

| Type | Magnitude of the main shock in a sequence | | | |
| --- | --- | --- | --- | --- |
| | $M \geq 5.5$ | 5.4–5.0 | 4.9–4.5 | $M \geq 4.5$ (total) |
| A  Swarm | 0 | 0 | 4 | 4 |
| B  Single | 0 | 2 | 12 | 14 |
| C  With foreshock | 0 | (1) | 0 | (1) |
| D  With foreshock and aftershock | 1 | 0 | 3 | 4 |
| E  With aftershock | 5 | 13 | 13 | 31 |
| Total | 6 | 16 | 32 | 54 |

district. The probability of giving an alarm in this case is therefore given by the ratio of 9 (sum of the A, C, D events) to 50 (the total number excluding swarms), i.e. an alarm rate $p_2 = 0.18$ is expected from the table. However, prediction will be successful only in the 5 events (C + D) of the 9 cases since swarm events are not the object of our prediction. In other words, the truth rate will be $p_1 = 0.56$.

These data illustrate the background of technical difficulty in prediction work. Fig. 8.11 shows a schematic view of the capability of earthquake prediction using the foreshock sequence method. Notice that strong earthquakes which are preceded by recognizable foreshock sequences occupy only a portion of the large group of seismic events under consideration. The hatched area in the figure represents earthquakes of

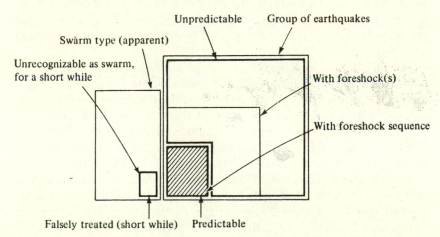

Fig. 8.11. Schematic diagram showing predictable and unpredictable earthquakes using the foreshock sequence method. (From Hsu, 1976.)

this type, which are considered to be predictable by the recognition of characteristic aspects of a precursory sequence. The area on the left represents swarm type events. The smallest part of this area represents especially confusing events, which resemble precursory seismic sequences in several respects, and might lead us to false predictions. Sooner or later, however, these are distinguished as swarms and the prediction cancelled, since some characteristic swarm aspects will appear in a later period of swarm activity (Hsu, 1976).

To summarize, certain types of earthquakes seem to be predictable, although they occupy, at present, only a small portion of the large group of disastrous earthquakes to be predicted. Types of predictable earthquakes may differ from place to place, depending on the seismotectonic environment. Even in a typical prediction by foreshock sequence, such as in the case of the Haicheng earthquake, prediction must be complemented by as many other kinds of information as possible to allow better recognition of precursors. Continuous comprehensive monitoring and interpretation of crustal activities are, therefore, essential.

### 8.5   *Further problems*

A remarkable aspect of earthquake prediction work is its dual constitution – on the one hand, as a basic science deeply concerned with the nature of earthquakes, and on the other, as an element in the reduction of earthquake hazards. In the latter sense, prediction is no longer a single scientific problem, but a new comprehensive discipline, which is related to various sociological problems. This is apparent in the discussion of the merit factor (§8.4), for example. If a numerical estimate of the merit factor is required, we are certainly concerned with complex socio-economic problems in order that the damage and cost parameters can be deduced correctly.

The question of how warnings can be most effectively made to the public, on which very little work has been done to the present time, is especially urgent. For example, how should earthquake predictions be released and utilized in such a way that the social, economic, legal and political effects are constructive and not counterproductive. What is the optimum level of preparation for emergency; and how should long-term planning for disaster mitigation be established? The importance of systematic investigation of this kind of problem has been emphasized by many workers concerned with earthquake prediction and hazards reduction (see for example Panel on the Public Policy Implications of Earthquake Prediction, 1975; Rikitake, 1976; Haas & Mileti, 1977).

A positive application of earthquake seismology to hazards reduction is the artificial control or modification of a potential earthquake source so that the energy may be allowed to dissipate gradually. Pioneering work at Denver, Colorado (see for example Evans, 1966), has shown the possibility that seismic energy may be dissipated safely by piecemeal fault slippings due to injection of water between the fracture surfaces. Perhaps the future may bring not only accurate earthquake prediction but also some control of earthquakes.

# Appendixes

Appendix 1. *Earthquake source parameters*

| Event | Date | $M$ | $M_S$ | $M_W$ | $M_o$ ($10^{27}$ dyn cm) | $L$ (km) | $D$ (km) | Slip type | $U$ (m) | $T$ (s) | $T^*$ (s) | $v_r$ (km s$^{-1}$) | $\Delta\sigma$ (bar) | $E_s$ ($10^{22}$ erg) | $E_f$ ($10^{22}$ erg) | Remarks |
|---|---|---|---|---|---|---|---|---|---|---|---|---|---|---|---|---|
| San Francisco | 18 Apr. 1906 | | 8.25 | 7.9 | 10 | 430 | 15 | RS | 5–7 | 7 | 10 | | | 200 | 300 | |
| Kanto | 1 Sep. 1923 | 7.9 | 8.2 | 7.9 | 7.6 | 130 | 70 | RT | 2.1 | | | | 21 | | | * |
| | | | | | | 85 | 55 | RT | 6.7 | | | | | | | |
| Tango | 27 Mar. 1927 | 7.5 | 7.75 | | 0.46 | 35 | 13 | LS | 3 | 6 | 2.5 | 2.3 | 115 | 10 | 4 | |
| North Izu | 25 Nov. 1930 | 7.0 | 7.1 | | 0.2 | 20 | 11 | LS | 3 | | 1.7 | | 150 | 2 | 3 | |
| Saitama | 21 Sep. 1933 | 7.0 | 6.75 | | 0.068 | 20 | 10 | LS | 1 | 2 | 1.6 | 2.3 | 59 | | | |
| Sanriku | 2 Mar. 1933 | 8.3 | 8.3 | 8.4 | 43 | 185 | 100 | N | 3.3 | 7 | 12 | 3.2 | 42 | | | |
| Long Beach | 11 Mar. 1933 | | 6.25 | | 0.028 | 30 | 15 | RS | 0.2 | 2 | 2.5 | 2.3 | 7 | | | |
| Imperial Valley | 19 May 1940 | | 7.1 | | 0.48 | 70 | 11 | RS | 2 | | 3.2 | | 55 | 1 | 1.5 | |
| Tottori | 10 Sep. 1943 | 7.4 | 7.4 | | 0.36 | 33 | 13 | RS | 2.5 | 3 | 4.0 | 2.3 | 99 | | | |
| Tonankai | 7 Dec. 1944 | 7.0 | 8.2 | 8.1 | 15 | 120 | 80 | T | 3.1 | | 9.2 | | 39 | | | |
| Mikawa | 12 Jan. 1945 | 7.1 | 7.1 | | 0.087 | 12 | 11 | RT | 2.2 | | 1.3 | | 140 | | | |
| Nankai | 20 Dec. 1946 | 8.1 | 8.2 | 8.1 | 15 | 120 | 80 | T | 3.1 | | 9.2 | | 39 | | | |
| | | | | | | 300 | 70–120 | RT | 4–6 | | | | | | | * |
| | | | | | | 320 | 50–140 | T | 5–18 | | | | | | | * |
| Fukui | 28 Jan. 1948 | 7.3 | 7.3 | | 0.33 | 30 | 13 | LS | 2 | 2 | 1.9 | 2.3 | 100 | | | |
| Tokachi-Oki | 4 Mar. 1952 | 8.1 | 8.3 | 8.1 | 17 | 180 | 100 | T | 1.9 | | 14 | | 17 | | | |
| Kern County | 21 Jul. 1952 | | 7.7 | | 2 | 60 | 18 | LT | 4.6 | 1 | 3.6 | | 140 | | | |
| Fairview | 16 Dec. 1954 | 7.1 | 7.1 | | 0.13 | 36 | 6 | RN | 2 | | 1.7 | | 100 | 6 | 7 | |
| Chile | 22 May 1960 | 8.3 | 8.3 | 9.5 | 2400 | 800 | 200 | T | 21 | 36 | 3.5 | 91 | | | * |
| | | | | | | 1000 | 120 | T | 20 | | | | | | | |
| Kitamino | 19 Aug. 1961 | 7.0 | 7.0 | | 0.09 | 12 | 10 | RT | 2.5 | 2 | 1.3 | 3.0 | 170 | | | |
| Wasaka Bay | 27 Mar. 1963 | 6.9 | 6.9 | | 0.033 | 20 | 8 | RS | 0.6 | 2 | 1.5 | 2.3 | 40 | | | |
| North Atlantic I | 3 Aug. 1963 | | 6.7 | | 0.12 | 32 | 11 | RS | 1 | | 2.2 | | 44 | | | |
| Kurile Islands | 13 Oct. 1963 | | 8.2 | 8.5 | 75 | 250 | 140 | T | 3 | | 17 | 3.5 | 28 | | | |
| North Atlantic II | 17 Nov. 1963 | | 6.5 | | 0.038 | 27 | 9 | RS | 0.48 | | 1.8 | | 24 | | | |
| Spain | 15 Mar. 1964 | | 7.1 | | 0.13 | 95 | 10 | T | 0.42 | | 3.6 | 1.4 | 11 | | | * |
| Alaska | 28 Mar. 1964 | | 8.5 | 9.2 | 520 | 500 | 300 | LT | 7 | 35 | 3.5 | 22 | 300 | 1000 | * |
| | | | | | | 600 | 200 | LT | 16 | | | | | | | |
| | | | | | | 800 | 175–290 | LT | 20 | | | | | | | |

| Location | Date | $M$ | $M_s$ | $M_w$ | $M_o$ | $L$ | $D$ | Slip type | $U$ | $T$ | $T^*$ | $t_r$ | $\Delta\sigma$ | $E_f$ |
|---|---|---|---|---|---|---|---|---|---|---|---|---|---|---|
| Niigata | 16 Jun. 1964 | 7.5 | 7.4 | | 3.2 | 80 | 30 | T | 3.3 | | 5.3 | 4.0 | 66 | 50 |
| Rat Island I | 4 Feb. 1965 | | 7.9 | 8.7 | 140 | 500 | 150 | T | 2.5 | | 25 | | 17 | 11 |
| Rat Island II | 30 Mar. 1965 | | 7.5 | | 3.4 | 50 | 80 | N | 1.2 | | 5.8 | 2.7 | 33 | |
| Parkfield | 28 Jun. 1966 | | 6.4 | | 0.032 | 26 | 7 | RS | 0.6 | 0.7 | 1.6 | | 32 | |
| Aleutian | 4 Jul. 1966 | | 7.2 | | 0.226 | 35 | 12 | RS | 1.6 | | 2.4 | | 64 | |
| Truckee | 12 Sep. 1966 | | 5.9 | | 0.0083 | 10 | 10 | RS | 0.3 | | 1.2 | | 20 | |
| Peru | 17 Oct. 1966 | | 7.5 | 8.1 | 20 | 80 | 140 | T | 2.6 | | 9.6 | | 41 | |
| Turkey | 22 Jul. 1967 | | 7.1 | | 0.83 | 80 | 20 | RS | 1.7 | | 4.7 | | 32 | |
| Borrego | 9 Apr. 1968 | | 6.7 | | 0.063 | 33 | 11 | RS | 0.58 | | 2.2 | | 22 | |
| Tokachi-Oki | 16 May 1968 | 7.9 | 8.0 | 8.2 | 28 | 150 | 100 | RT | 4.1 | | 12 | 3.5 | 37 | |
| Saitama | 1 Jul. 1968 | 6.1 | 5.8 | | 0.019 | 10 | 6 | LS | 0.92 | 1 | 0.9 | 3.4 | 100 | |
| Iran | 31 Aug. 1968 | | 7.3 | | 1 | 80 | 20 | T | 2.1 | | 4.7 | | 38 | |
| Portuguese | 28 Feb. 1969 | | 8.0 | | 5.5 | 80 | 50 | T | 2.5 | | 6.1 | | 53 | |
| Kurile Islands | 11 Aug. 1969 | | 7.8 | 8.2 | 22 | 180 | 85 | LS | 2.9 | | 12 | 3.5 | 28 | |
| Gifu | 9 Sep. 1969 | 6.6 | 6.6 | | 0.035 | 18 | 10 | LS | 0.6 | 1 | 1.7 | 2.5 | 35 | |
| Peru | 31 May 1970 | | 7.8 | 7.9 | 10 | 130 | 70 | N | 1.6 | 1 | 8.7 | 2.5 | 28 | |
| San Fernando | 9 Feb. 1971 | | 6.6 | | 0.12 | 20 | 14 | LT | 1.4 | 1 | 2.0 | 2.4 | 62 | |
| Nemuro-Oki | 17 Jun. 1973 | 7.4 | 7.7 | | 6.7 | 60 | 100 | T | 1.6 | | 7.5 | | 35 | |
| China | 27 Jul. 1976 | | 8.0 | 7.5 | 2 | 150 | 15 | RS | 2.7 | | | | | |

Symbols:
$M$   magnitude from local data, e.g. magnitude in 'Rika-nenpyo' (*Annual Table of Scientific Constants*; Maruzen Publishing Co., Tokyo) for the Japanese earthquakes;
$M_s$   surface wave magnitude;
$M_w$   magnitude calculated from the seismic moment (see Kanamori, 1977);
$M_o$   seismic moment in $10^{27}$ dyn cm;
$L$   fault length (km);
$D$   fault depth (definition should be slightly changed for a buried fault) (km);
Slip type   N, normal; T, reverse; S, strike-slip; RS, right-lateral; LS, left-lateral; RT, reverse with right-lateral; LT, reverse with left-lateral;
$U$   final slip (average) (m);
$T$   rise time (linear ramp time function) (s);
$T^*$   theoretical rise time (see Geller, 1976) (s);
$t_r$   rupture velocity (km s$^{-1}$);
$\Delta\sigma$   stress-drop (bar);
$E_s$   energy of seismic waves (approximate) ($10^{22}$ erg);
$E_f$   strain-energy change in faulting (approximate) ($10^{22}$ erg);
*   geodetic model, principally related to land movements.

The majority of the data are taken from Geller (1976). Further information and references to the original events may be obtained from this paper as well as from Kanamori (1977).

Appendix 2. *Earthquake magnitude, M, versus the associated surface faulting and deformation* (*empirical relations*)

| Parameter | Unit | Formula (reference) | Region |
|---|---|---|---|
| $L$: fault length (typical) | km | $\log L = 1.02\ M - 5.76$ (Tocher, 1958) | USA |
| | | $\log L = 1.32\ M - 7.99$ (Iida, 1965) | World-wide |
| | | $\log L = 1.14\ M - 6.38$ (Ambraseys & Zatopek, 1968) | Anatolia |
| | | $\log L = 0.6\ M - 2.9$ (Matsuda, 1975) | Japan |
| $r$: radius of deformation area | km | $\log r = 0.51\ M - 2.27$ (Dambara, 1966) | Japan |
| $L_m$: fault length (maximum observed) | km | $\log L_m = 0.5\ M - 1.75$ (Iida, 1965) | World-wide |
| | | $\log L_m = 0.5\ M - 1.8$ (Otsuka, 1964) | World-wide |
| | | $\log L_m = 0.35\ M - 0.26$ (Bonilla, 1970) | USA |
| | | $\log L_m = 0.5\ M - 1.9$ (Yonekura, 1972) | Japan |
| $U$: fault offset (typical) | m | $\log U = 0.55\ M - 3.71$ (Iida, 1965) | World-wide |
| | | $\log U = 0.96\ M - 6.69$ (Chinnery, 1969) | World-wide (strike-slip) |
| | | $\log U = 0.57\ M - 3.91$ (Bonilla, 1970) | USA |
| | | $\log U = 0.6\ M - 4.0$ (Matsuda, 1975) | Japan |
| $U_m$: fault offset (maximum observed) | m | $\log U_m = 0.57\ M - 3.19$ (Bonilla, 1970) | USA |
| | | $\log U_m = 0.67\ M - 4.33$ (Yonekura, 1972) | Japan |

Compiled by Matsuda (1976).

Appendix 3. *WWSSN station codes and locations*

| Code | Station | Location |
|------|---------|----------|
| Fig. 3.14 | | |
| ADE | Adelaide | South Australia, Australia |
| AFI | Afiamalu | Samoa Islands |
| BAG | Baguio City | Luzon, Philippines |
| BEC | Bermuda | Bermuda |
| BKC | Bald Knob | Arkansas, USA |
| BKS | Byerly | California, USA |
| CTA | Charters Towers | Queensland, Australia |
| GDH | Godhavn | Greenland |
| GOL | Golden | Colorado, USA |
| HKC | Hong Kong | Hong Kong |
| HNR | Honiara | Solomon Islands |
| IST | Istanbul | Turkey |
| KIP | Kipapa Oahu | Hawaii, USA |
| KTG | Kap Tobin | Greenland |
| MAN | Manila (W) | Luzon, Philippines |
| MDS | Madison | Wisconsin, USA |
| MUN | Mundaring | Western Australia, Australia |
| NAI | Nairobi | Kenya |
| NUR | Nurmijarvi | Finland |
| PRE | Pretoria | Transvaal, South Africa |
| RAB | Rabaul | New Britain |
| SHI | Shiraz | Iran (Persia) |
| TOL | Toledo | Spain |
| TUC | Tucson | Arizona, USA |
| | | |
| Fig. 5.10 | | |
| ARE | Arequipa | Peru |
| ESK | Eskdalemuir | Scotland, UK |
| COL | College Outpost | Alaska and Aleutians, USA |
| KIP | Kipapa Oahu | Hawaii, USA |
| LPB | La Paz | Bolivia |
| MAT | Matsushiro | Nagano, Honshu, Japan |
| NAT | Natal | Rio Grande do Norte, Brazil |
| TRN | Trinidad (W) | Trinidad and Tobago |

# References

## Chapter 1

Kasahara, K. (1969) Focal processes and various approaches to their mechanism. In *A Symposium on Processes in the Focal Region* (eds. K. Kasahara & A. E. Stevens), Publ. Dominion Obs., Ottawa, Vol. 37, pp. 187–9.

Kasahara, K. (1971) The role of geodesy in crustal movement studies. In Recent Crustal Movements (eds. B. W. Collins & R. Fraser), *Bull. Roy. Soc. New Zealand*, **9**, 1–5.

Matuzawa, T. (1964) *Study of Earthquakes*, Uno Shoten, Tokyo, pp. 1–213.

Simon, R. B. (1968) *Earthquake Interpretations*, Colorado School of Mines, Colorado.

## Chapter 2

Aki, K. (1956) Some problems in statistical seismology. *Zisin, J. Seismol. Soc. Japan*, **8**, 205–28 (in Japanese).

Båth, M. & Duda, S. J. (1964) Earthquake volume, fault plane area, seismic energy, strain, deformation, and related quantities. *Ann. Geofis.* (Rome), **17**, 353–68.

Dambara, T. (1966) Vertical movements of the earth's crust in relation to the Matsushiro earthquake. *J. Geod. Soc. Japan*, **12**, 18–45 (in Japanese).

Gutenberg, B. (1945) Magnitude determination for deep-focus earthquakes. *Bull. Seismol. Soc. Am.*, **35**, 117–30.

Gutenberg, B. & Richter, C. F. (1942) Earthquake magnitude, intensity, energy and acceleration. *Bull. Seismol. Soc. Am.*, **32**, 163–91.

Gutenberg, B. & Richter, C. F. (1944) Frequency of earthquakes in California. *Bull. Seismol. Soc. Am.*, **34**, 185–8.

Gutenberg, B. & Richter, C. F. (1954) *Seismicity of the Earth and Associated Phenomena*, 2nd edn, Princeton University Press, Princeton, New Jersey.

Gutenberg, B. & Richter, C. F. (1956a) Earthquake magnitude, intensity, energy and acceleration (second paper). *Bull. Seismol. Soc. Am.*, **46**, 105–45.

Gutenberg, B. & Richter, C. F. (1956b) Magnitude and energy of earthquakes. *Ann. Geofis.* (Rome), **9**, 1–15.

Iida, K. (1959) Earthquake energy and earthquake fault. *J. Earth Sci., Nagoya University*, **7**, 98–107.

Iida, K. (1965) Earthquake magnitude, earthquake fault, and source dimensions. *J. Earth Sci., Nagoya University*, **13**, 115–32.

Kasahara, K. (1957) The nature of seismic origins as inferred from seismological and geodetic observations (1). *Bull. Earthq. Res. Inst., Tokyo University*, **35**, 473–532.

Lomnitz, C. (1974) *Global Tectonics and Earthquake Risk*, Developments in Geotectonics Series, Vol. 5, Elsevier Sci., Amsterdam.

Otsuka, M. (1965) Earthquake magnitude and surface fault formation. *Zisin, J. Seismol. Soc. Japan*, **18**, 1–8 (in Japanese).

Press, F. (1967) Dimensions of the source region for small shallow earthquakes. In *Proceedings, VESIAC Conference, Shallow Source Mechanisms, VESIAC Report*, **7885-1-X**, 155–63.

Richter, C. F. (1935) An instrumental earthquake magnitude scale. *Bull. Seismol. Soc. Am.*, **25**, 1–32.

Richter, C. F. (1958) *Elementary Seismology*, W. H. Freeman & Co., San Francisco.

Rikitake, T. (1976) *Earthquake Prediction*, Developments in Geotectonics Series, Vol. 9, Elsevier Sci., Amsterdam.

Terashima, T. (1968) Magnitude of microearthquakes and the spectra of microearthquake waves. *Bull. Int. Inst. Seismol. Earthq. Eng.*, **5**, 31–108.

Tocher, D. (1958) Earthquake energy and ground breakage. *Bull. Seismol. Soc. Am.*, **48**, 147–52.

Tsuboi, C. (1933) Investigation on the deformation of the earth's crust found by precise geodetic means. *Jap. J. Astron. Geophys.*, **10**, 93–248.

Tsuboi, C. (1956) Earthquake energy, earthquake volume, aftershock area, and strength of the earth's crust. *J. Phys. Earthq.*, **4**, 63–6.

Tsuboi, C. (1965) Time rate of earthquake energy release in and near Japan. *Proc. Jap. Acad.*, **41**, 392–7.

Utsu, T. & Seki, A. (1955) A relation between the area of after-shock region and the energy of main shock. *Zisin, J. Seismol. Soc. Japan*, **7**, 233–40 (in Japanese).

## Chapter 3

Aki, K. (1960a) The use of Love waves for the study of earthquake mechanism. *J. Geophys. Res.*, **65**, 323–31.

Aki, K. (1960b) Study of earthquake mechanism by a method of phase equalization applied to Rayleigh and Love waves. *J. Geophys. Res.*, **65**, 729–40.

Balakina, L. M., Savarensky, E. F. & Vvedenskaya, A. V. (1961) On determination of earthquake mechanism. *Physics and Chemistry of the Earth* (eds. L. H. Ahres *et al.*), **4**, 211–38.

Båth, M. (1974) *Spectral Analysis in Geophysics*, Developments in Solid Earth Geophysics Series, Vol. 7, Elsevier Sci., Amsterdam.

Bessonova, E. N., Gotsadze, O. D., Keilis-Borok, V. I., *et al.* (1960) Investigation of the mechanism of earthquakes. *Soviet Res. Geophys.* (English translation), **4**, 1–201.

Bullen, K. E. (1953) *An Introduction to the Theory of Seismology*, 2nd edn, Cambridge University Press, Cambridge, pp. 1–296.

Byerly, P. (1955) Nature of faulting as deduced from seismograms. In *Crust of the Earth* (Geol. Soc. Am. special paper 62), pp. 75–85.

Dillinger, W. H., Pope, A. J. & Harding, S. T. (1971). The determination of focal mechanisms using P- and S-wave data. *NOAA Technical Report* No. 44, *pp.* 1–56.

Hirasawa, T. (1966) A least square method for the focal mechanism determination from S-wave data; Part I. *Bull. Earthq. Res. Inst., Tokyo University*, **44**, 901–18.

Hodgson, J. H. (ed.) (1959) *The Mechanics of Faulting, with Special Reference to the Fault-Plane Work* (*A Symposium*), Publ. Dominion Obs., Ottawa, Vol. 20, pp. 253–418.

Hodgson, J. H. (ed.) (1961) *A Symposium on Earthquake Mechanism*. Publ. Dominion Obs., Ottawa, Vol. 24, pp. 301–97.

Hodgson, J. H. & Stevens, A. E. (1964) Seismicity and earthquake mechanism. In *Research in Geophysics*, Vol. 2, *Solid Earth and Interface Phenomena* (ed. H. Odishaw), MIT Press, Cambridge, Mass., pp. 27–59.

Hodgson, J. H. & Storey, R. S. (1953) Tables extending Byerly's fault-plane technique to earthquakes of any focal depth. *Bull Seismol. Soc. Am.*, **43**, 49–61.

Honda, H. (1957) The mechanism of the earthquakes. *Sci. Reports, Tohoku University*, Series 5, *Geophysics*, **9**, 1–46. Reprinted in *The Mechanics of Faulting, with Special Reference to the Fault-Plane Work (A Symposium)* (ed. J. H. Hodgson), Publ. Dominion Obs., Ottawa, Vol. 20 (1959).

Honda, H. (1962) Earthquake mechanism and seismic waves. *Geophys. Notes, Faculty Sci., Tokyo University*, **15** (supplement), 1–97.

Ichikawa, M. (1971) Reanalyses of mechanism of earthquakes which occurred in and near Japan, and statistical studies on the nodal plane solutions obtained, 1926–1968. *Geophys. Mag., JMA*, **35**, 207–74.

Japan Meteorological Agency (1971) *A Manual on Seismometrical Work (Data Interpretation)*. Japan Meteorological Agency, Tokyo (in Japanese).

Kanamori, H. (1970) Synthesis of long-period surface waves and its application to earthquake source studies – Kurile Islands earthquake of October 13, 1963. *J. Geophys. Res.*, **75**, 5011–27.

Kasahara, K. (1963) Radiation mode of S waves from a deep-focus earthquake as derived from observations. *Bull. Seismol. Soc. Am.*, **53**, 643–59.

Kasahara, K. & Stevens, A. E. (eds.) (1969) *A Symposium on Processes in the Focal Region*. Publ. Dominion Obs., Ottawa Vol. 37, pp. 183–235.

Kawasumi, H. (1937) An historical sketch of the development of knowledge concerning the initial motion of an earthquake. *Publ. Bur. Cent. Seism. Int.*, A**15**, 258–330.

Keilis-Borok, V. I. (1950) Concerning the determination of the seismic parameters of a focus. *Tr. Geofiz. Inst. Akad. Nauk. SSSR*, **9**, 3–19 (in Russian).

Keilis-Borok, V. I. (1959) The study of earthquake mechanism. In *The Mechanics of Faulting, with Special Reference to the Fault-Plane Work (A Symposium)* (ed. J. H. Hodgson), Publ. Dominion Obs., Ottawa, Vol. 20, p. 279.

Knopoff, L. (1961) Analytical calculation of the fault-plane problem. In *A Symposium on Earthquake Mechanism* (ed. J. H. Hodgson) Publ. Dominion Obs., Ottawa, Vol. 24, pp. 309–15.

Love, A. E. H. (1944) *A Treatise on the Mathematical Theory of Elasticity*, 4th edn, Dover, New York.

Maruyama, T. (1968) *Basic Theory of Seismic Waves*. Part I of *Earthquakes, Volcanoes, and Rockmechanics* (ed. S. Miyamura), Kyoritsu Shuppan Co., Tokyo (in Japanese).

Nakano, H. (1923) Notes on the nature of the forces which give rise to the earthquake motions. *Seismol. Bull., Central Meteorological Obs., Japan*, **1**, 92–120.

Nuttli, O. & Whitmore, J. D. (1962) On the determination of the polarization angle of the S wave. *Bull. Seismol. Soc. Am.*, **52**, 95–107.

Ritsema, A. R. (1959) On the focal mechanism of southeast Asian earthquakes. In *The Mechanics of Faulting, with Special Reference to the Fault-Plane Work (A Symposium)* (ed. J. H. Hodgson), Publ. Dominion Obs., Ottawa, Vol. 20, pp. 253–418.

Stauder, W. (1962) The focal mechanism of earthquakes. *Advances in Geophysics* (eds. H. E. Landsberg & J. Van Mieghem), **9**, 1–76.

Stefánsson, R. (1966) The use of transverse waves in focal mechanism studies. *Tectonophysics*, **3**, 35–60.

Stevens, A. E. (1967) S.-wave earthquake mechanisms equations. *Bull. Seismol. Soc. Am.*, **57**, 99–112.

Sykes, L. R. (1967) Mechanism of earthquakes and nature of faulting on the mid-oceanic ridges. *J. Geophys. Res.*, **72**, 2131–53.

Udias, A. (1964) A least squares method for earthquake mechanism determination using S-wave data. *Bull. Seismol. Soc. Am.*, **54**, 2037–47.

Wickens, A. J. & Hodgson, J. H. (1967) Computer re-evaluation of earthquake mechanism solutions, 1922–1962. Publ. Dominion Obs., Ottawa, Vol. 33, pp. 1–560.

Wilson, J. T. (1965) A new class of faults and their bearing on continental drift. *Nature*, **207**, 343–7.

## Chapter 4

Abe, K. (1973) Tsunami and mechanism of great earthquakes. *Phys. Earth Planet. Interiors*, **7**, 143–53.

Ando, M. (1971) A fault-origin model of the great Kanto earthquake of 1923 as deduced from geodetic data. *Bull. Earthq. Res. Inst., Tokyo University*, **49**, 19–32.

Burgers, J. M. (1939) Some considerations on the fields of stress connected with dislocations in a regular crystal lattice, I. *Proc. Koninkl. Ned. Acad. Wetenschap.*, **42**, 293–325.

Bilby, B. A. & Eshelby, J. D. (1968) Dislocations and the theory of fracture. In *Fracture, An Advanced Treatise* (ed. H. Liebowitz), Academic Press, New York, pp. 99–182.

Bullen, K. E. (1953) *An Introduction to the Theory of Seismology*, 2nd edn, Cambridge University Press, Cambridge, pp. 1–296.

Chinnery, M. A. (1961) The deformation of the ground around surface faults. *Bull. Seismol. Soc. Am.*, **51**, 355–72.

Chinnery, M. A. (1963) The stress changes that accompany strike-slip faulting. *Bull. Seismol. Soc. Am.*, **53**, 921–32.

Chinnery, M. A. (1967) Theoretical fault models. In *Symposium, Processes in the Focal Region* (eds. K. Kasahara & A. E. Stevens), Publ. Dominion Obs., Ottawa, Vol. 37, pp. 211–23.

Dieterich, J. H. (1974) Earthquake mechanisms and modeling. *Ann. Review, Earth Planet. Sci.*, **2**, 275–301.

Eshelby, J. D. (1957) The determination of the elastic field of an ellipsoidal inclusion, and related problems. *Proc. Roy. Soc. London*, **A241**, 376–96.

Hastie, L. M. & Savage, J. C. (1970) A dislocation model for the 1964 Alaska earthquake (Letter). *Bull Seismol. Soc. Am.*, **60**, 1389–92.

Honda, H. (1932). On the mechanism and the types of the seismograms of shallow earthquakes. *Geophys. Mag. JMA*, **5**, 69–88.

Hoshino, K. (1956) On magnitude of earthquakes accompanied by faults (Letter) *Zisin, J. Seismol. Soc. Japan*, 2nd Series, **8**, 160–62 (in Japanese).

Iacopi, R. (1976) *Earthquake country*, 5th edn, Lane Books, Menlo Park, California, pp. 1–160.

Iida, K. (1959) Earthquake energy and earthquake fault. *J. Earth Sci., Nagoya University*, **7**, 98–107.

Iida, K. (1965) Earthquake magnitude, earthquake fault and source dimensions. *J. Earth Sci., Nagoya University*, **13**, 115–32.

Kanamori, H. (1973) Mode of strain release associated with major earthquakes in Japan. *Ann. Review, Earth Planet. Sci.*, **1**, 213–39.

Kanamori, H. & Ando, M. (1973) Fault parameters of the Great Kanto earthquake of 1923. In *Publications for the 50th Anniversary of the Great Kanto Earthquake, 1923*, Earthq. Res. Inst., Tokyo University, 1 September 1973, pp. 89–101 (in Japanese with English abstract).

Kasahara, K. (1957) The nature of seismic origins as inferred from seismological and geodetic observations (1). *Bull. Earthq. Res. Inst., Tokyo University*, **35**, 473–532.

Kasahara, K. (1958*a*) The nature of seismic origins as inferred from seismological and geodetic observations (2). *Bull. Earthq. Res. Inst., Tokyo University*, **36**, 21–53.

Kasahara, K. (1958*b*) Physical conditions of earthquake faults as deduced from geodetic data. *Bull. Earthq. Res. Inst., Tokyo University*, **36**, 455–64.

Kasahara, K. (1960) Static and dynamic characteristics of earthquake faults, Earthquake Research Institute, The University of Tokyo, pp. 74–5.

Kasahara, K. (1967) Focal processes and various approaches to their mechanism. In *Symposium, Processes in the Focal Region* (eds. K. Kasahara and A. E. Stevens), Publ. Dominion Obs., Ottawa, Vol. 37, pp. 187–9.

Knopoff, L. (1958) Energy release in earthquakes. *Geophys. J. MNRAS*, **1**, 44–52.

Land Survey Department (1930) *Result, Revision 3rd Order Triangulations Tango District*, Land Survey Department, Tokyo, pp. 1–8 (in Japanese).

Landau, L. D. & Lifshitz, E. M. (1970) *Theory of Elasticity*, 2nd edn, Pergamon, Oxford.

Love, A. E. H. (1944) *A Treatise on the Mathematical Theory of Elasticity*, 4th edn, Dover, New York.

Mansinha, L. & Smylie, D. E. (1971). The displacement fields of inclined faults. *Bull. Seismol. Soc. Am.*, **61**, 1433–40.

Maruyama, T. (1963) On the force equivalents of dynamical elastic dislocations with reference to the earthquake mechanism. *Bull. Earthq. Res. Inst., Tokyo University*, **41**, 46–86.

Maruyama, T. (1964) Statical elastic dislocation in an infinite and semi-infinite medium. *Bull. Earthq. Res. Inst., Tokyo University*, **42**, 289–368.

Maruyama, T. (1966) On two-dimensional elastic dislocations in an infinite and semi-infinite medium. *Bull. Earthq. Res. Inst., Tokyo University*, **44**, 811–71.

Maruyama, T. (1973) Theoretical model of seismic faults. In *Publications for the 50th Anniversary of the Great Kanto Earthquake, 1923*, Earthq. Res. Inst., Tokyo University, 1 September, 1973, pp. 147–65 (in Japanese with English abstract).

Matsu'ura, M. & Sato, R. (1975). Displacement fields due to the fault. *Zisin, J. Seismol. Soc. Japan*, **28**, 429–34 (in Japanese with English abstract).

Otsuka, M. (1965) Earthquake magnitude and surface fault formation. *Zisin, J. Seismol. Soc. Japan*, 2nd Series, **18**, 1–8 (in Japanese with English abstract).

Plafker, G. (1972) Alaskan earthquake of 1964 and Chilean earthquake of 1960: implications for arc tectonics. *J. Geophys. Res.*, **77**, 901–25.

Press, F. (1965) Displacements, strains, and tilts at teleseismic distances. *J. Geophys. Res.*, **70**, 2395–412.,

Press, F. (1967) Dimensions of the source region for small shallow earthquakes. In *Proceedings, VESIAC Conference, Shallow Source Mechanisms. VESIAC Report*, **7885-1-X**, 155–63.

Reid, H. F. (1911) The elastic-rebound theory of earthquakes. *University of California Publ. Geol. Sci.*, **6**, 413–44.

Savage, J. C. & Hastie, L. M. (1966) Surface deformation associated with dip-slip faulting. *J. Geophys. Res.*, **71**, 4897–904.

Starr, A. T. (1928) Slip in a crystal and rupture in a solid due to shear. *Proc. Camb. Phil. Soc.*, **24**, 489–500.

Steketee, J. A. (1958*a*) On Volterra's dislocations in a semi-infinite elastic medium. *Can. J. Phys.*, **36**, 192–205.

Steketee, J. A. (1958*b*) Some geophysical applications of the elasticity theory of dislocations. *Can. J. Phys.*, **36**, 1168–98.

Thomson, R. M. & Seitz, F. (1968) Surface of solids. In *Fracture, An Advanced Treatise* (ed. H. Liebowitz), Academic Press, New York, pp. 1–97.

Tocher, D. (1958) Earthquake energy and ground breakage. *Bull. Seismol. Soc. Am.*, **48**, 147–52.

Tsuboi, C. (1933) Investigation on the deformation of the earth's crust found by precise geodetic means. *Jap. J. Astron. Geophys.*, **10**, 93–248.

Volterra, V. (1907) Sur l'équilibre des corps élastiques multiplement connexes. *Ann. Sci. Ecole Norm. Super., Paris*, **24**, 401–517.

Wyss, M. & Brune, J. N. (1968) Seismic moment, stress, and source dimensions for earthquakes in California–Nevada region. *J. Geophys. Res.*, **73**, 4681–94.

## Chapter 5

Abe, K. (1973) Tsunami and mechanism of great earthquakes. *Phys. Earth Planet. Interiors*, **7**, 143–53.

Aki, K. (1966*a*) Generation and propagation of G waves from the Niigata earthquake of June 16, 1964. Part 1. A statistical analysis. *Bull. Earthq. Res. Inst., Tokyo University*, **44**, 23–72.

Aki, K. (1966*b*) Generation and propagation of G waves from the Niigata earthquake of June 16, 1964. Part 2. Estimation of earthquake moment, released energy, and stress-strain drop from the G-wave spectrum. *Bull. Earthq. Res. Inst., Tokyo University*, **44**, 73–88.

Aki, K. (1968) Seismic displacements near a fault. *J. Geophys. Res.*, **73**, 5359–75.

Allen, C. R. & Smith, S. W. (1966). Parkfield earthquake of June 27–29, 1966. Pre-earthquake and post-earthquake surficial displacement. *Bull. Seismol. Soc. Am.*, **56**, 966–7.

Båth, M. (1974) *Spectral Analysis in Geophysics*, Developments in Solid Earth Geophysics Series, Vol. 7, Elsevier Sci., Amsterdam.

Benioff, H., Press, F. & Smith, S. W. (1961) Excitation of the free oscillations of the earth by earthquakes. *J. Geophys. Res.*, **66**, 605–18.

Ben-Menahem, A. (1961) Radiation of seismic surface-waves from finite moving sources. *Bull. Seismol. Soc. Am.*, **51**, 401–35.

Ben-Menahem, A. & Toksöz, M. N. (1963) Source mechanism from spectra of long-period seismic surface waves. 3. The Alaska earthquake of July 10, 1958. *Bull. Seismol. Soc. Am.*, **53**, 905–19.

Bessonova, E. N., Gotsadze, O. D. & Keilis-Borok, V. I. *et al.* (1960) Investigation of the mechanism of earthquakes. *Soviet Res. Geophys.*, **4**. (English translation, American Geophysical Union, Consultants Bureau, New York.)

Brown, R. D., Jr. & Vedder, J. G. (1967) Surface tectonic fractures along the San Andreas fault. The Parkfield–Cholame, California, earthquakes of June–August 1966. *US Geol. Surv. Profess. Paper*, **579**.

Brune, J. N. (1970) Tectonic stress and the spectra of seismic shear waves from earthquakes. *J. Geophys. Res.*, **75**, 4997–5009.

Bullen, K. E. (1953) *An Introduction to the Theory of Seismology*, 2nd edn, Cambridge University Press, Cambridge, pp. 1–296.

Burridge, R. & Knopoff, L. (1964) Body force equivalence for seismic dislocations. *Bull. Seismol. Soc. Am.*, **54**, 1875–88.

Burridge, R. & Knopoff, L. (1967) Model and theoretical seismicity. *Bull. Seismol. Soc. Am.*, **57**, 341–71.

Douglas, B. M. & Ryall, A. (1972) Spectral characteristics and stress drop for microearthquakes near Fairview Peak, Nevada. *J. Geophys. Res.*, **77**, 351–9.

Eaton, J. P. (1967) The Parkfield–Cholame, California, earthquake of June–August 1966 – Instrumental seismic studies. *US Geol. Surv. Profess. Paper*, **579**, 57–65.

Fukao, Y. (1979) Tsunami earthquakes and subduction processes near deep-sea trenches. *J. Geophys. Res.*, **84**, 2303–14.

Hanks, T. C. & Thatcher, W. (1972) A graphical representation of seismic source parameters. *J. Geophys. Res.*, **77**, 4393–405.

Hanks, T. C. & Wyss, M. (1972) The use of body-wave spectra in the determination of seismic-source parameters. *Bull. Seismol. Soc. Am.*, **62**, 561–89.

Hasegawa, H. S. (1974) Theoretical synthesis and analysis of strong motion spectra of earthquakes. *Canadian Geotech. J.*, **11**, 278–97.

Haskell, N. A. (1964*a*) Radiation pattern of surface waves from point sources in a multi-layered medium. *Bull. Seismol. Soc. Am.*, **54**, 377–93.

Haskell, N. A. (1964*b*) Total energy and energy spectral density of elastic wave radiation from propagating faults. *Bull. Seismol. Soc. Am.*, **54**, 1811–41.

Haskell, N. A. (1969) Elastic displacements in the near-field of a propagating fault. *Bull. Seismol. Soc. Am.*, **59**, 865–908.

Hirasawa, T. (1965) Source mechanism of the Niigata earthquake of June 16, 1964, as derived from body waves. *J. Phys. Earth*, **13**, 35–66.

Hirasawa, T. & Stauder, W. (1965) On the seismic body waves from a finite moving source. *Bull. Seismol. Soc. Am.*, **55**, 237–62.

Ishida, M. (1974) Determination of fault parameters of small earthquakes in the Kii Peninsula, *J. Phys. Earth*, **22**, 177–212.

Kanamori, H. (1970) Synthesis of long-period surface waves and its application to earthquake source studies – Kurile Islands earthquake of October 13, 1963. *J. Geophys Res.*, **75**, 5011–27.

Kanamori, H. (1972*a*) Determination of effective tectonic stress associated with earthquake faulting. The Tottori earthquake of 1943. *Phys. Earth Planet. Interiors*, **5**, 426–34.

Kanamori, H. (1972*b*) Mechanism of tsunami earthquakes. *Phys. Earth Planet. Interiors*, **6**, 346–59.

Kanamori, H. (1973) Mode of strain release associated with major earthquakes in Japan. *Ann. Review, Earth Planet. Sci.*, **1**, 213–39.

Kanamori, H. (1974). A new view of earthquakes. In *Physics of the Earth* (*A Modern View of the Earth*) (ed. Physical Society of Japan), Maruzen, Tokyo, pp. 261–82 (in Japanese).

Knopoff, L. & Gilbert, F. (1959) Radiation from a strike-slip fault. *Bull. Seismol. Soc. Am.*, **49**, 163–78.

Love, A. E. H. (1944) *A Treatise on the Mathematical Theory of Elasticity*, 4th edn, Dover, New York.

Maruyama, T. (1963) On the force equivalents of dynamical elastic dislocations with reference to the earthquake mechanism. *Bull. Earthq. Res. Inst., Tokyo University*, **41**, 467–86.

Mogi, A., Kawamura, B. & Iwabuchi, Y. (1964) Submarine crustal movement due to the Niigata earthquake in 1964, in the environs of the Awa Sima Island, Japan Sea. *J. Geod. Soc. Japan,* **10**, 180–6.

Nakamura, K., Kasahara, K. & Matsuda, T. (1964) Tilting and uplift of an Island, Awashima, near the epicentre of the Niigata earthquake in 1964. *J. Geod. Soc. Japan,* **10**, 172–86.

Press, F., Ben-Menahem, A. & Toksöz, M. N. (1961). Experimental determination of earthquake fault length and rupture velocity. *J. Geophys. Res.,* **66**, 3471–85.

Savage, J. C. (1972) Relation of corner frequency to fault dimensions. *J. Geophys. Res.,* **77**, 3788–95.

Savage, J. C. & Mansinha, L. (1963) Radiation from a tensile fracture. *J. Geophys. Res.,* **68**, 6345–58.

Starr, A. T. (1928) Slip in a crystal and rupture in a solid due to shear. *Proc. Camb. Phil. Soc.,* **24**, 489–500.


**Chapter 6**

Aki, K. (1967) Scaling law of seismic spectrum. *J. Geophys. Res.,* **72**, 1217–31.

Aki, K. (1972) Scaling law of earthquake source time-function. *Geophys. J. MNRAS,* **31**, 3–25.

Anderson, L. D. & Whitcomb, J. H. (1973) The dilatancy–diffusion model of earthquake prediction. In *Proceedings, Conference on Tectonic Problems of the San Andreas Fault System* (eds. R. L. Kovach & A Nur), Stanford University Publ., Geol. Sci. Vol. 13, p. 417.

Båth, M. & Duda, S. J. (1964) Earthquake volume, fault plane area, seismic energy, strain, deformation and related quantities. *Ann. Geofis* (Rome), **17**, 353–68.

Benioff, H. (1951) Earthquakes and rock creep. *Bull. Seismol. Soc. Am.,* **41**, 31–62.

Benioff, H. (1964) Earthquake source mechanisms. *Science,* **143**, 1399–406.

Brace, W. F. (1968) Current laboratory studies pertaining to earthquake prediction. *Tectonophysics,* **6**, 75–87.

Brace, W. F. (1969) The mechanical effects of pore pressure on fracture of rocks. *Geol. Surv. Can. Pap.,* **68–52**, 113–23.

Brace, W. F. & Byerlee, J. D. (1966) Stick-slip as a mechanism of earthquakes. *Science,* **153**, 990–2.

Brace, W. F. & Byerlee, J. D. (1970) California earthquakes: why only shallow focus? *Science,* **168**, 1573–5.

Brace, W. F., Pauling, B. W., Jr. & Scholz, C. H. (1966). Dilatancy in the fracture of crystalline rocks. *J. Geophys. Res.,* **71**, 3939–53.

Brady, B. T. (1974) Theory of earthquakes, 1. A scale independent theory of rock failure. *Pure Appl. Geophys.,* **112**, 701–25.

Bridgman, P. (1949) Volume changes in the plastic stages of simple compression. *J. Appl. Phys.,* **20**, 1241.

Brune, J. N. (1970) Tectonic stress and the spectra of seismic shear waves from earthquakes. *J. Geophys. Res.,* **75**, 4997–5009.

Brune, J. N., Henyey, T. L. & Roy, R. F. (1969) Heat flow, stress and rate of slip along the San Andreas fault. *J. Geophys. Res.,* **74**, 3821–7.

Byerlee, J. D. (1967) Frictional characteristics of granite under high confining pressure. *J. Geophys. Res.,* **72**, 3639–48.

Byerlee, J. D. (1968) Brittle–ductile transition in rocks. *J. Geophys. Res.,* **73**, 4741–50.

Byerlee, J. D. (1970a) Static and kinetic friction of granite at high normal stress. *Int. J. Rock Mech. Mineral. Sci.*, **7**, 577–82.

Byerlee, J. D. (1970b) The mechanism of stick-slip. *Tectonophysics*, **9**, 475–86.

Byerlee, J. D. & Brace, W. F. (1969) High-pressure mechanical instability in rocks. *Science*, **164**, 713–5.

Dennis, J. G. & Walker, C. T. (1965) Earthquakes resulting from metastable phase transition. *Tectonophysics*, **2**, 401–7.

Dieterich, J. H. (1974) Earthquake mechanisms and modeling. *Ann. Review, Earth Planet. Sci.*, **2**, 275–301.

Eshelby, J. D. (1957) The determination of the elastic field of an ellipsoidal inclusion and related problems. *Proc. Roy. Soc. London*, **A241**, 376–96.

Evison, F. F. (1963) Earthquakes and faults. *Bull. Seismol. Soc. Am.*, **53**, 873–91.

Griffith, A. A. (1920) The phenomena of rupture and flow in solids. *Phil. Trans. Roy. Soc. London*, **A221**, 163–98.

Griggs, D. T. & Baker, D. W. (1969) The origin of deep-focus earthquakes. In *Properties of Matter under Unusual Conditions* (eds. H. Mark & S. Fernbach) Wiley Interscience, New York. pp. 23–42.

Griggs, D. T. & Handin, J. (1960) Observations on fracture and a hypothesis of earthquakes. *Geol. Soc. Am. Memoirs*, **79**, 347–64.

Griggs, D. T. & Handin, J. (1966). *Handbook of Physical Constants*, Geol. Soc. Am. Memoirs, Vol. 97.

Honda, H. (1957) The mechanism of the earthquakes. *Sci. Reports, Tohoku University*, Series 5, *Geophysics*, **9**, 1–46.

Jeffreys, H. (1936) Note on fracture. *Proc. Roy. Soc. Edinburgh*, **A56**, 158–63.

Kanamori, H. (1977) The energy release in great earthquakes. *J. Geophys. Res.*, **82**, 2981–7.

Kanamori, H. & Anderson, D. L. (1975) Theoretical basis of some empirical relations in seismology. *Bull. Seismol. Soc. Am.*, **65**, 1073–95.

Liebowitz, H. (ed.) (1968) *Fracture, An Advanced Treatise*, Vol. 1, Academic Press, New York, pp. 1–597.

McGarr, A. (1977) Seismic moments of earthquakes beneath island arcs, phase changes, and subduction velocities. *J. Geophys. Res.*, **82**, 256–62.

Mjachkin, V. I., Brace, W. F., Sobolev, G. A. & Dieterich, J. H. (1975) Two models for earthquake forerunners. *Pure Appl. Geophys.*, **113**, 169–81.

Mogi, K. (1966) Pressure dependence of rock strength and transition from brittle fracture to ductile flow. *Bull. Earthq. Res. Inst., Tokyo University*, **44**, 215–32.

Mogi, K. (1967) Earthquakes and fractures. *Tectonophysics*, **5**, 35–55.

Mogi, K. (1974) Rock fracture and earthquake prediction. *J. Soc. Materials Sci. Japan*, **23**, 320–31 (in Japanese).

Mott, N. F. (1948) Fracture of metals: theoretical considerations. *Engineering*, **165**, 16.

Nur, A. (1972) Dilatancy, pore fluids, and premonitory variations of $t_s/t_p$ travel times. *Bull. Seismol. Soc. Am.*, **62**, 1217.

Ohnaka, M. (1976) A physical basis for earthquakes based on the elastic rebound model. *Bull. Seismol. Soc. Am.*, **66**, 433–51.

Orowan, E. (1950) In *Fatigue and Fracture of Metals* (ed. W. M. Murray), Wiley, New York, pp. 139–57.

Orowan, E. (1960) Mechanism of seismic faulting. *Geol. Soc. Am. Memoirs*, **79**, 323–45.

Orowan, E. (1966) Dilatancy and the seismic focal mechanism. *Rev. Geophys.*, **4**, 395–404.

Raleigh, C. B. (1967) Tectonic implications of serpentinite weakening. *Geophys. J. MNRAS*, **14**, 113–8.

Raleigh, C. B. & Paterson, M. S. (1965) Experimental deformation of serpentinite and its tectonic implications. *J. Geophys. Res.*, **70**, 3965–85.

Randall, M. J. & Knopoff, L. (1970) The mechanism at the focus of deep earthquakes. *J. Geophys. Res.*, **75**, 4965–76.

Reid, H. F. (1911) The elastic-rebound theory of earthquakes. *University of California Publ. Geol. Sci.*, **6**, 413–44.

Richter, C. F. (1958) *Elementary Seismology*, W. H. Freeman & Co., San Francisco, pp. 1–768.

Rikitake, T. (1976) *Earthquake Prediction*, Developments in Solid Earth Geophysics Series, Vol. 9, Elsevier Sci., Amsterdam.

Ringwood, A. E. (1972) Phase transformations and mantle dynamics. *Earth Planet. Sci. Letts.*, **14**, 233–41.

Sato, T. & Hirasawa, T. (1973) Body wave spectra from propagating shear cracks. *J. Phys. Earth*, **21**, 415–31.

Savage, J. C. (1969) The mechanics of deep-focus faulting. *Tectonophysics*, **8**, 115–27.

Savage, J. C. & Wood, M. D. (1971) The relation between apparent stress and stress drop. *Bull. Seismol. Soc. Am.*, **61**, 1381–8.

Scholz, C. H. (1968) Microfracturing and the inelastic deformation of rock in compression, *J. Geophys. Res.*, **73**, 1417–32.

Scholz, C. H., Molnar, P. & Johnson, T. (1972) Detailed studies of frictional sliding of granite and implications for the earthquake mechanisms. *J. Geophys. Res.*, **77**. 6392–406.

Scholz, C. H., Sykes, L. R. & Aggarwal, Y. P. (1973) Earthquake prediction: A physical basis. *Science*, **181**, 803.

Smith, R. B., Winkler, P. L., Anderson, J. G. & Scholz, C. H. (1974) Source mechanisms of microearthquakes associated with underground mines in eastern Utah. *Bull. Seismol. Soc. Am.*, **64**, 1295–317.

Stuart, W. S. (1974) Diffusionless dilatancy model for earthquake precursors. *Geophys. Res. Letts.*, **1**, 261.

Sykes, L. R. (1968) Deep earthquakes and rapidly-running phase changes, a reply to Dennis and Walker. *J. Geophys. Res.*, **73**, 1508–10.

Takeuchi, H. & Kikuchi, M. (1973) A dynamical model of crack propagation. *J. Phys. Earth*, **21**, 27–37.

Wyss, M. (ed.) (1975) Earthquake prediction and rock mechanics. *Pure Appl. Geophys.*, **113**, 1–330.

Wyss, M. & Brune, J. N. (1968) Seismic moment, stress, and source dimensions for earthquakes in California–Nevada region. *J. Geophys. Res.*, **73**, 4681–94.

Yamashita, T. (1976) On the dynamical process of fault motion in the presence of friction and inhomogeneous initial stress. Part I. Rupture propagation. *J. Phys. Earth*, **24**, 417–44.

**Chapter 7**

Allen, C. R. (1968) The tectonic environments of seismically active and inactive areas along the San Andreas fault system. *Proceedings, Conference on Geologic Problems of San Andreas Fault System* (eds. W. R. Dickinson & A. Grantz), Stanford University Publ., Geol. Sci. Vol. 11, pp. 70–82.

Allen, C. R. (1969) Active faulting in northern Turkey. Contr. No. 1577, Div. Geol. Sci., Calif. Inst. Tech., p. 32.

Allen, C. R., St Amand, P., Richter, C. F. & Nordquist, J. M. (1965) Relationship between seismicity and geologic structure in the southern California region. *Bull. Seismol. Soc. Am.*, **55**, 753–97.

Anderson, D. L. (1962). The plastic layer of the Earth's mantle. *Scientific American*, **207**, 52–9.

Anderson, D. L. (1975) Accelerated plate tectonics. *Science*, **187**, 1077–9.

Aoki, H. (1977) Possibility of a great earthquake in Tokai district. In *A Symposium on Earthquake Prediction Research* (eds. Z. Suzuki & S. Omote), Seismological Society of Japan, Tokyo, pp. 56–68 (in Japanese).

Atwater, T. (1970) Implications of plate tectonics for the Cenozoic tectonic evolution of western North America. *Bull. Geol. Soc. Am.*, **81**, 3513–36.

Bird, J. M. & Isacks, B. (eds.) (1972) *Plate Tectonics* (*Selected Papers from the Journal of Geophysical Research*), American Geophysical Union, Washington, DC, pp. 1–563.

Bott, M. H. P. & Dean, D. S. (1973) Stress diffusion from plate boundaries. *Nature*, **243**, 339–41.

Brown, R. D., Jr & Wallace, R. E. (1968) Current and historic fault movement along the San Andreas fault between Paicines and Camp Dix, California. *Proceedings, Conference on Geologic Problems of San Andreas Fault System* (eds. W. R. Dickinson & A. Grantz), Stanford University Publ., Geol. Sci. Vol. 11, pp. 22–41.

Brune, J. N. (1968) Seismic moment, seismicity, and rate of slip along major fault zones. *J Geophys. Res.*, **73**, 777–84.

Bufe, C. G., Harsh, P. W. & Burford, R. O. (1977) Steady-state seismic slip – a precise recurrence model. *Geophys. Res. Letts.*, **4**, 91–4.

Clarke, S. H., Jr & Nilsen, T. H. (1973) Displacement of Eocene strata and implications for the history of offset along the San Andreas Fault, central and northern California. In *Proceedings, Conference on Tectonic Problems of San Andreas Fault System* (eds. R. L. Kovach & A. Nur), Stanford University Publ., Geol. Sci. Vol. 13, pp. 358–367.

Dickinson, W. R. & Grantz, A. (eds.) (1968) *Proceedings of Conference on Geologic Problems of San Andreas Fault System*, Stanford University Publ., Geol. Sci. Vol. 11, pp. 1–374.

Fedotov, S. A. (1963) The absorption of transverse seismic waves in the upper mantle and energy classification of near earthquakes of immediate focal depth. *Izv. Akad. Nauk SSSR* (*Ser. Geophys.*) (English translation), **6**, 509–20.

Fedotov, S. A. (1965) Regularities of the distribution of strong earthquakes in Kamchatka, the Kurile Islands and northeastern Japan. *Acad. Sci. USSR Trudy Inst. Phys. Earth*, **36**, 66–93.

Grantz, A. & Dickinson, W. R. (1968) Indicated cumulative offsets along the San Andreas fault in the California Coast Ranges. In *Proceedings, Conference on Geologic Problems of San Andreas Fault System* (eds. W. R. Dickinson & A. Grantz), Stanford University Publ., Geol. Sci. Vol. 11, pp. 117–20.

Gutenberg, B. & Richter, C. F. (1954) *Seismicity of the Earth and Associated Phenomena*, 2nd edn, Princeton University Press, Princeton, New Jersey.

Harada, T. & Kassai, A. (1971) Horizontal strain of the crust in Japan for the last 60 years. *J. Geod. Soc. Japan*, **17**, 4–7 (in Japanese).

Heiskanen, W. A. & Vening Meinesz, F. A. (1958) *The Earth and its Gravity Field*, McGraw-Hill Book Co., New York.

Hill, M. L. & Dibblee, T. W., Jr (1953) San Andreas, Garlock, and Big Pine faults, California and southwestern Oregon. *US Geol. Survey Profess. Paper*, **501-c**, 1–9.

Ichikawa, M. (1966) Statistical investigation of mechanism of earthquakes occurring in and near Japan and some related problems. *J. Meteorol. Res., JMA,* **18**, 83–154 (in Japanese).

Ida, Y. (1974) Slow-moving deformation pulses along tectonic faults. *Phys. Earth Planet. Interiors,* **9**, 328–37.

Isacks, B., Oliver, J. & Sykes, L. R. (1968) Seismology and the new global tectonics. *J. Geophys. Res.,* **73**, 5855–99.

Ishii, H. (1977) Characteristics of crustal movement observed at wide area. In *A Symposium on Earthquake Prediction Research* (eds. Z. Suzuki & S. Omote), Seismological Society of Japan, Tokyo, pp. 116–26 (in Japanese).

Kanamori, H. (1971) Great earthquakes at island arcs and the lithosphere. *Tectonophysics,* **12**, 187–98.

Kanamori, H. (1973) Mode of strain release associated with major earthquakes in Japan. *Ann. Review, Earth Planet. Sci.,* **1**, 213–39.

Kanamori, H. (1977) Seismic and aseismic slip along subduction zones and their tectonic implications. In *Island Arcs, Deep Sea Trenches and Back Arc Basins* (eds. M. Talwani & W. C. Pitman III), American Geophysical Union, pp. 163–74.

Kanamori, H. & Anderson, D. L. (1975) Theoretical basis of some empirical relations in seismology. *Bull. Seismol. Soc. Am.,* **65**, 1073–95.

Kasahara, K. (1971). The role of geodesy in crustal movement studies. In Recent Crustal Movements (eds. B. W. Collins & R. Fraser), *Bull. Roy. Soc. New Zealand,* **9**, 1–5.

Kasahara, K. (1973a) Earthquake fault studies in Japan. *Phil. Trans. Roy. Soc. London,* **A274**, 287–96.

Kasahara, K. (1973b) Tiltmeter observation in complement with precise levellings. *J. Geod. Soc. Japan,* **19**, 93–9 (in Japanese).

Kasahara, K. (1975) Aseismic faulting following the 1973 Nemuro-oki earthquake, Hokkaido, Japan (a possibility). In Special Issue: Earthquake Prediction and Rock Mechanics (ed. M. Wyss), *Pure Appl. Geophys.,* **113**, 127–39.

Kasahara, K. (1979) Migration of crustal deformation. *Tectonophysics,* **52**, 329–41.

Kasahara, K. & Sugimura, A. (1964) Spatial distribution of horizontal secular strain in Japan. *J. Geod. Soc. Japan,* **10**, 139–45.

Kelleher, J. A. (1972) Rupture zones of large South American earthquakes and some predictions. *J. Geophys. Res.,* **77**, 2087–103.

King, C. Y., Nason, R. D. & Tocher, D. (1973) Kinematics of fault creep. *Phil. Trans. Roy. Soc. London,* **A274**, 355–60.

Kovach, R. L. & Nur, A. (eds.) (1973) *Proceedings, Conference on Tectonic Problems of San Andreas Fault System,* Stanford University Publ., Geol. Sci. Vol. 13, pp. 1–494.

Lensen, G. J. (1965) Active faults and major earthquakes in New Zealand. *New Zealand J. Geol. Geophys.,* **8**, No. 6 (pull-out between pp. 900 and 901).

Le Pichon, X. (1968) Sea-floor spreading and continental drift. *J. Geophys. Res.,* **73**, 3661–97.

Le Pichon, X., Francheteau, J. & Bonnin, J. (1973) *Plate Tectonics,* Developments in Solid Earth Geophysics Series, Vol. 7, Elsevier Sci., Amsterdam.

Lomnitz, C. (1974) *Global Tectonics and Earthquake Risk,* Developments in Solid Earth Geophysics Series, Vol. 5, Elsevier Sci., Amsterdam.

Love, A. E. H. (1944) *A Treatise on the Mathematical Theory of Elasticity,* 4th edn, Dover, New York.

Matsuda, T. (1967) Earthquake geology. In Special Issue: Seismology in Japan, *Zisin*, J. *Seismol. Soc. Japan*, **20**, 230–5 (in Japanese).

Matsuda, T. (1975) Magnitude and recurrence interval of earthquakes from a fault. *Zisin*, J. *Seismol. Soc. Japan*, **28**, 269–83 (in Japanese).

Matsuda, T. (1976) Active faults and earthquakes – the geological aspect. *Memoir, Geol. Soc. Japan*, **12**, 15–32 (in Japanese).

Matsuda, T. (1977) Active faults and preestimation of earthquakes. In *Symposium on Earthquake Prediction Research* (eds. Z. Suzuki & S. Omote) Seismological Society of Japan, Tokyo, pp. 194–202 (in Japanese).

Matsuda, T., Okada, A. & Huzita, K. (1976) Active faults in Japan. *Memoir, Geol. Soc. Japan*, **12**, 185–98 (in Japanese).

McKenzie, D. P. & Parker, R. L. (1967) The North Pacific: an example of tectonics on a sphere. *Nature*, **216**, 1276–80.

Miyabe, N. (1955) Vertical earth movement in Nankai district. *Bull. Geogr. Survey Inst.*, **4**, 1–14.

Mogi, K. (1968a) Migration of seismic activity. *Bull. Earthq. Res. Inst., Tokyo University*, **46**, 53–74.

Mogi, K. (1968b) Development of aftershock areas of great earthquakes. *Bull. Earthq. Res. Inst., Tokyo University*, **46**, 175–203.

Mogi, K. (1968c) Sequential occurrence of recent great earthquakes. *J. Phys. Earth*, **16**, 30–6.

Mogi, K. (1969) Some features of recent seismic activity in and near Japan. (2) Activity before and after great earthquakes. *Bull. Earthq. Res. Inst., Tokyo University*, **47**, 395–417.

Mogi, K. (1973) Relationship between shallow and deep seismicity in the western Pacific region. *Tectonophysics*, **17**, 1–22.

Mogi, K. (1977) Seismic activity and earthquake prediction. In *A Symposium on Earthquake Prediction Research* (eds. Z. Suzuki & S. Omote), Seismological Society of Japan, Tokyo, pp. 203–14 (in Japanese).

Nakamura, K. (1977) Why do Pacific coastal terraces on peninsulas tilt away from trenches? *Special Issue, Tokai Section, Coordinating Com. Earthq. Prediction*. Geographical Survey Institute, Tokyo, pp. 41–44 (in Japanese).

Nakamura, K., Kasahara, K. & Matsuda, T. (1964) Tilting and uplift of an island, Awashima, near the epicentre of the Niigata earthquake in 1964. *J. Geod. Soc. Japan*, **10**, 172–9.

Nason, R. & Weertman, J. (1973) A dislocation theory analysis of fault creep events. *J. Geophys. Res.*, **78**, 7745–51.

Okada, A. & Nagata, T. (1953) Land deformation of the neighborhood of Muroto point after the Nankaido great earthquake in 1946. *Bull. Earthq. Res. Inst., Tokyo University*, **31**, 169–77.

Oliver, J. & Isacks, B. (1967) Deep earthquake zones, anomalous structures in the upper mantle, and the lithosphere. *J. Geophys. Res.*, **72**, 4259–75.

Richter, C. F. (1958) *Elementary Seismology*, W. H. Freeman & Co., San Francisco, pp. 611–16.

Research Group for Crustal Movement, Tohoku University (1977) Analysis of crustal movement in the Tohoku district observed by array system. In Proceedings, Symposium on Recent Crustal Movements (ed. I. Nakagawa), *J. Geod. Soc. Japan*, **22**, 225–318.

Savage, J. C. (1971) A theory of creep waves propagating along a transform fault. *J. Geophys. Res.*, **76**, 1954–66.

Savage, J. C. & Burford, R. O. (1973) Geodetic determination of relative plate motion in central California. *J. Geophys. Res.*, **78**, 832–45.

Scholz, C. H. (1977) A physical interpretation of the Haicheng earthquake prediction. *Nature*, **267**, 121–4.

Shimazaki, K. (1977) A model of earthquake recurrence and its application to crustal movement in Tokai district, Japan. In *Report, Tokai Division, Coordinating Com. Earthquake Prediction*. Geographical Survey Institute, Tokyo. pp. 32–40 (in Japanese).

Stein, R. S., Thatcher, W. & Castle, O. (1977) Initiation and development of the southern California uplift along its northern margin. Presented at the 1977 *Recent Crustal Movement Symposium*, Palo Alto, July, 1977.

Sugimura, A. & Uyeda, S. (1973) *Island Arcs, Japan and its Environs*. Developments in Geophysics Series, Vol. 3, Elsevier Sci., Amsterdam.

Sykes, L. R. (1967) Mechanism of earthquakes and nature of faulting on the mid-oceanic ridges. *J. Geophys. Res.*, **72**, 2131–51.

Tanaka, Y., Otsuka, S. & Lazo, L. (1977) Migrating crustal deformations in Peru. *Abstract, Annual Meeting, Seismological Society of Japan*, 1977, No. 2, p. 82.

Tocher, D. (1960) Creep rate and related measurement at Vineyard, California. *Bull. Seismol. Soc. Am.*, **50**, 396–404.

Utsu, T. (1966) Regional differences in absorption of seismic waves in the upper mantle as inferred from abnormal distributions of seismic intensities. *J. Fac. Sci., Hokkaido University, Japan*, Series VII, **2(4)**, 359–74.

Utsu, T. (1972) Large earthquakes near Hokkaido and the expectancy of the occurrence of a large earthquake off Nemuro. In *Report, Coordinating Com. Earthquake Prediction*, Geographical Survey Institute, Tokyo, Vol. 7, pp. 7–13 (in Japanese).

Utsu, T. (1976a) Historical major earthquakes off Tokaido, Honshu, Japan. In *Special Issue, Section Report, Coordinating Com. Earthquake Prediction*, Geographical Survey Institute, Tokyo, Vol. 1, pp. 1–8 (in Japanese).

Utsu, T. (1976b) Problems on predicting earthquakes in East Hokkaido. In *Special Issue, Section Report, Coordinating Com. Earthquake Prediction*, Geographical Survey Institute, Tokyo, Vol. 1, pp. 45–64 (in Japanese).

Whitcomb, J. H., Allen, C. R., Garmany, J. D. & Hileman, J. A. (1973) Focal mechanisms and tectonics. San Fernando Earthquake Series. 1971, *Rev. Geophys. Space Phys.*, **11**, 693–730.

Whitten, C. A. (1956) Crustal movement in California and Nevada. *Am. Geophys. Union Trans.*, **37**, 398–8.

Wilson, J. T. (1965) A new class of faults and their bearing on continental drift. *Nature*, **207**, 343–7.

Yamada, J. (1973) A water-tube tiltmeter and its application to crustal movement studies. *Report, Earthq. Res. Inst.*, **10**, 1–147 (in Japanese).

Yoshii, T. (1973) Upper mantle structure beneath the north Pacific and the marginal seas. *J. Phys. Earth*, **21**, 313–28.

Yoshikawa, T., Kaizuka, S. & Ota, Y. (1964) Crustal movement in the late Quaternary revealed with coastal terraces on the southeast coast of Shikoku, southwestern Japan. *J. Geod. Soc. Japan*, **10**, 116–22.

## Chapter 8

Aggarwal, Y. P., Sykes, L. R., Simpson, D. W. & Richards, P. G. (1975) Spatial and temporal variations in $t_s/t_p$ and in P-wave residuals at Blue Mountain Lake: application to earthquake prediction. *J. Geophys. Res.*, **80**, 718–32.

Aki, K. (1956) Some problems in statistical seismology. *Zisin, J. Seismol. Soc. Japan*, **8**, 205–228 (in Japanese).

Allen, C. R., St Amand, P., Richter, C. F. & Nordquist, J. M. (1965) Relationship between seismicity and geologic structure in the southern California region. *Bull. Seismol. Soc. Am.*, **55**, 753–97.

Båth, M. (1974) *Spectral Analysis in Geophysics*, Developments in Solid Earth Geophysics Series, Vol. 7, Elsevier Sci., Amsterdam.

Chu, F. M. (1976) Review of prediction, warning and disaster prevention in the Haicheng earthquake (*M* = 7.3). *Proceedings of the Lectures by the Seismological Delegation of the People's Republic of China.* Seismological Society of Japan, Tokyo, pp. 15–26 (in Japanese).

Clark, M. M., Grantz, A. & Rubin, M. (1972) Holocene activity of the Coyote Creek fault as recorded in sediments of Lake Cahuilla. *US Geol. Surv. Profess. Paper*, **787**, 112–30.

Dambara, T. (1973) Crustal movements before, at and after the Niigata earthquake. *Report, Coordinating Com. Earthquake Prediction.* Geographical Survey Institute, Tokyo, Vol. 9, pp. 93–6 (in Japanese).

Earthquake Disaster Prevention Society (1977) *Precursors of Big Earthquakes* (Posthumous work of A. Imamura), Kokin Shoin, Tokyo, pp. 1–170 (in Japanese).

Evans, D. M. (1966) Man-made earthquakes in Denver. *Geotimes*, **10**, 11–18.

Haas, J. E. & Mileti, P. S. (1977) Socioeconomic and political consequences of earthquake prediction. In the US–Japan Seminar on Theoretical and Experimental Investigations of Earthquake Precursors (eds. Z. Suzuki & C. Kisslinger), *J. Phys. Earth*, **25** (supplement), 285–93.

Hsu, S. H. (1976) Characteristics of seismic activities in the Haicheng earthquake, *Proceedings of the Lectures by the Seismological Delegation of the People's Republic of China.* Seismological Society of Japan, Tokyo, pp. 27–41 (in Japanese).

Imamura, A. (1937) *Theoretical and Applied Seismology*, Maruzen, Tokyo.

Ishii, H. (1976) Application of prediction method for analysis of crustal movement. In Symposium, Recent Crustal Movements (ed. I. Nakagawa), *J. Geod. Soc. Japan*, **22**, 299–301.

Kakimi, T., Sato, H., Tsumura, K. & Ishida, M. (1977). Seismicity, crustal movements and Neotectonics in Kanto district. *A Symposium on Earthquake Prediction Research* (eds. 2. Suzuki & S. Omote), Seismological Society of Japan, Tokyo, pp. 21–45 (in Japanese).

Kanamori, H. & Chung, W. Y. (1974) Temporal changes in P-wave velocity in southern California. *Tectonophysics*, **23**, 67–78.

Kasahara, K. (1973) Tiltmeter observation in complement with precise levellings. *J. Geod. Soc. Japan*, **19**, 93–9 (in Japanese).

Kawasumi, H. (1951) Measures of earthquake danger and expectancy of maximum intensity throughout Japan as inferred from the seismic activity in historical times. *Bull. Earthq. Res. Inst., Tokyo University*, **29**, 469–82.

Kawasumi, H. (1970) Proofs of 69 years periodicity and imminence of destructive earthquake in southern Kwanto district and problems in the countermeasures thereof. *Chigaku Zasshi*, **76**, 115–138 (in Japanese).

Kuno, H. (1962) The old and new Tanna tunnels. *Kagaku*, **32**, 397–401 (in Japanese).

Lensen, G. J. (1968). Analysis of progressive fault displacement during downcutting at the Branch river terraces, South Island, New Zealand. *Bull. Geol. Soc. Am.*, **79**, 545–56.

Lomnitz, C. (1974) *Global Tectonics and Earthquake Risk*, Developments in Geotectonics Series, Vol. 5, Elsevier Sci., Amsterdam.

Matsuda, T. (1975a) Magnitude and recurrence interval of earthquakes from a fault. *Zisin, J. Seismol. Soc. Japan*, **28**, 269–83 (in Japanese).

Matsuda, T. (1975b) Active fault assessment for Irozati fault system, Izu Peninsula. In *Report on the Earthquake off the Izu Peninsula, 1974 and the Disaster* (ed. R. Tsuchi), pp. 121–5 (in Japanese).

Matsuda, T. (1976) Active faults and earthquakes – the geological aspect. *Memoir, Geol. Soc. Japan*, **12**, 15–32 (in Japanese).

Matsuda, T. (1977) Active faults and preestimation of earthquakes. In *A Symposium on Earthquake Prediction Research* (eds. Z. Suzuki & S. Omote). Seismological Society of Japan, Tokyo, pp. 194–202 (in Japanese).

Mogi, K. (1968) Migration of seismic activity. *Bull. Earthq. Res. Inst., Tokyo University*, **46**, 53–74.

Okada, A. (1973) On the Quaternary faulting along the Median Tectonic Line. In *Median Tectonic Line* (ed R. Sugiyama), Tokai University Press, Tokyo, pp. 49–86 (in Japanese).

Panel on the Public Policy Implications of Earthquake Prediction (1975) *Earthquake Prediction and Public Policy*, National Academy of Science, Washington, DC, pp. 1–142.

Press. F., Benioff. H., Frosch. R. A., Griggs, D. T., Hadin, J., Hanson, R. E., Hess, H. H., Housner, G. W., Munk, W. H., Orowan, E., Pakiser Jr., L. C., Sutton, G. & Tocher, D. (1965) *Earthquake Prediction: a Proposal for a Ten Year Program of Research*, Office, Sci. Technol., Washington, DC, pp. 1–134.

Raleigh, B., Bennett, G., Craig, H., Hanks, T., Molnar, P., Nur, A., Savage, J., Scholz, C., Turner, R. & Wu, F. (1977) Prediction of the Haicheng Earthquake. *EOS, Trans. Am. Geophys. Union*, **58**, 236–72.

Rikitake, T. (1957a) Statistics of ultimate strain of the earth's crust and probability of earthquake occurrence. *Tectonophysics*, **26**, 1–21.

Rikitake, T. (1975b) Dilatancy model and empirical formulas for an earthquake area. *Pure Appl. Geophys.*, **113**, 141–7.

Rikitake, T. (1976) *Earthquake Prediction*, Developments in Solid Earth Geophysics Series, Vol. 9, Elsevier Sci., Amsterdam.

Rikitake, T. (1977) Possible procedure of earthquake prediction and some problems of earthquake warning. In *A Symposium on Earthquake Prediction Research* (eds. Z. Suzuki & S. Omote), Seismological Society of Japan, Tokyo, pp. 215–24 (in Japanese).

Sato, H. & Inouchi, N. (1977) On relations between the ground uplifts and the earthquakes. In *A Sympsium on Earthquake Prediction Research* (eds. Z. Suzuki, & S. Omote), Seismological Society of Japan, Tokyo, pp. 138–44 (in Japanese).

Scholz. C. H. (1977) A physical interpretation of the Haicheng earthquake prediction. *Nature*, **267**, 121–4.

Scholz, C. H., Sykes, L. R. & Aggarwal, Y. P. (1973) Earthquake prediction: a physical basis. *Science*, **181**, 803–9.

Semenov. A. I. (1969) Change of transversal and shear wave travel time ratio before strong earthquakes. *Izv. Acad. Sci. USSR*, **3**, 72–7 (in Russian).

Shimazaki, K. (1972) Hidden periodicities of destructive earthquakes at Tokyo. *Zisin, J. Seismol. Soc. Japan*, **25**, 24–32 (in Japanese).

Suzuki, Z. & Kisslinger, C. (eds.) (1977) Earthquake precursors. In Proc. US-Japan Seminar on Theoretical and Experimental Investigations of Earthquake Precursors. *J. Phys. Earth*, **25** (supplement), 1–296.

Tsuboi, C., Wadati, K. & Hagiwara, T. (1962) Prediction of earthquakes progress to date and plans for further development. *Report, Earthquake Prediction*

*Research Group, Japan.* Earthquake Research Institute, University of Tokyo, Tokyo, pp. 1–21 (in Japanese).

Tsubokawa, I. (1969) On relation between duration of crustal movement and magnitude of earthquake expected. *J. Geod. Soc. Japan*, **15**, 75–88 (in Japanese).

Tsubokawa, I. (1973) On relation between duration of precursory geophysical phenomena and duration of crustal movement before earthquake. *J. Geod. Soc. Japan*, **19**, 116–9 (in Japanese).

Tsubokawa, I., Ogawa, Y. & Hayashi. T. (1964) Crustal movements before and after the earthquake. *J. Geod. Soc. Japan*, **10**, 165–71.

Usami, T. & Hisamoto, S. (1970) Future probability of a coming earthquake with intensity V or more in the Tokyo area. *Bull. Earthq. Res. Inst., Tokyo University*, **48**, 331–40 (in Japanese).

Utsu, T. (1977) Probabilities in earthquake prediction. *Zisin, J. Seismol. Soc. Japan*, **30**, 179–85 (in Japanese).

Wallace, R. E. (1968) *Notes on Stream Channels Offset by the San Andreas Fault, Southern Coast Ranges, California.* Stanford University Publ., Geol. Sci. Vol. 11, pp. 6–21.

Wallace, R. E. (1970) Earthquake recurrence intervals on the San Andreas fault. *Bull. Geol. Soc. Am.*, **81**, 2875–90.

Whitcomb, J. H., Garmany, J. D. & Anderson, D. L. (1973) Earthquake prediction: variation of seismic velocities before the San Fernando Earthquake. *Science*, **180**, 632–5.

Wyss, M. (ed.) (1975). Earthquake Prediction and Rock Mechanics, Special Issue; *Pure Appl. Geophys.*, **113**, 1–330.

## Appendix 1

Geller, R. J. (1976) Scaling relations for earthquake source parameters and magnitudes. *Bull. Seismol. Soc. Am.*, **66**, 1501–23.

Kanamori, H. (1977) The energy release in great earthquakes. *J. Geophys. Res.*, **82**, 2981–7.

## Appendix 2

Ambraseys, N. N. & Zatopek, A. (1968) The Varto Ustukran (Anatolia) earthquake of 19 August, 1966. *Bull. Seismol. Soc. Am.*, **58**, 47–102.

Bonilla, M. G. (1970) Surface faulting and related effects. In *Earthquake Engineering* (ed. R. L. Wiegel), Prentice-Hall, Inc., New Jersey.

Chinnery, M. A. (1969) Earthquake magnitude and source parameters. *Bull. Seismol. Soc. Am.*, **59**, 1969–82.

Dambara, T. (1966) Vertical movements of the Earth's crust in relation to the Matsushiro earthquake, *J. Geod. Soc. Japan*, **12**, 18–45 (in Japanese with English abstract).

Iida, K. (1965) Earthquake magnitude, earthquake fault, and source dimensions. *J. Earth Sci., Nagoya Univ.*, **13**, 115–32.

Matsuda, T. (1975) Magnitude and recurrence interval of earthquakes from a fault. *Zisin, J. Seismol. Soc. Japan*, **28**, 269–83 (in Japanese with English abstract).

Matsuda, T. (1976) Active faults and earthquakes – the geological aspect. *Memoir, Geol. Soc. Japan*, **12**, 15–22 (in Japanese with English abstract).

Otsuka, M. (1964) Earthquake magnitude and surface fault formation. *J. Phys. Earth*, **12**, 19–24.

Tocher, D. (1958) Earthquake energy and ground breakage. *Bull. Seismol. Soc. Am.*, **48**, 147–53.

Yonekura, N. (1972) A review on seismic crustal deformations in and near Japan. *Bull. Dept Geogr., Univ. Tokyo*, **4**, 17–50.

# Author index

# Subject index